THE SKY
AT NIGHT

THE SKY
AT NIGHT

PATRICK MOORE

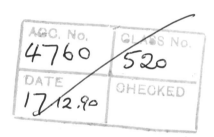

HARRAP
London

First published in Great Britain 1989
by HARRAP BOOKS Ltd
19–23 Ludgate Hill, London EC4M 7PD

ISBN 0 245–54731–2

Design by Gwyn Lewis

Typeset by
Poole Typesetting (Wessex) Ltd, Bournemouth
Printed in Great Britain by Butler and Tanner Ltd,
Frome, Somerset

Contents

Preface

This is the ninth volume in *The Sky at Night* series. I have followed the usual pattern, and though the chapters are based upon the actual television programmes I have brought them up to date where necessary. In trying to make each chapter self-contained I realize that I have been guilty of a certain amount of overlap, but there seemed to be no sensible alternative; I have also omitted some months, because where two programmes dealt with the same subject I have had to combine them. The period covered extends from the beginning of 1985 through to the autumn of 1988. It has certainly been an eventful time; it has seen the Voyager 2 pass of Uranus, the return of Halley's Comet, and the outburst of a supernova in the Large Cloud of Magellan, to mention only a few of the highlights. I hope, therefore, that this book also tells the story of what has happened over these three years.

I am most grateful not only to those who have appeared with me on the programmes (listed separately) but also to those who have allowed me to use their pictures; to all at Messrs. Harrap; to that splendid television producer Pieter Morpurgo, without whom the programmes could never have been made, and to Paul Doherty, for his outstanding illustrations which have been used in this book as well as in the programmes themselves.

Patrick Moore
Selsey, September 1988.

Acknowledgments

During the period covered in this book (January 1985 to August 1988) many guests have appeared with me in *Sky at Night* programmes. In chronological order, they are: David Malin; Dr Jim Emerson; Dr Lionel Wilson; Ron Arbour; Iain Nicolson; Dr Garry Hunt; Dr Edward Stone; Professor Carl Sagan; Professor Fred Whipple; Professor Jan Oort; Professor Raoul Sagdeev; Dr Rüdeger Reinhard; Professor Susan McKenna-Lawlor; Dr Tony McDonnell; Dr Alan Johnstone; Dr David Dale; Dr Paul Murdin; Professor Sir Francis Graham-Smith; Dr David Clark; Douglas Arnold; Andrew Murray; Dr Ron Maddison; Professor Alec Boksenberg; Dr Jasper Wall; Brian Mack; Dr Michael Morris; Stanley Atkinson, Briam Edwards; Dr Richard Hills; Dr Ron Newport; Dr Ian McLean; Dr Russell Cannon; Professor Sir Bernard Lovell; Dr Jim Cohen; Dr Ford Davies; Dr Richard Davies; Dr Tom Muxlow; Dudley Fuller; Dr Richard Dreiser; Dr Al Harper; Dr Kyle Cudworth; Dr W. W. Morgan; Dr Halton C. Arp; Dr John Mason; Dr David Latham, and Dr Bruce Campbell: To all these — my most grateful thanks.

The photograph on p. 75 of Halley's Comet is Copyright © 1986 Royal Observatory, Edinburgh and that of Rho Ophiuchi on p. 184 is Copyright © 1980 Royal Observatory, Edinburgh and Anglo-Australian Telescope Board, and I am grateful for permission to reproduce these photographs. The photograph on p. 108 of part of the Large Magellanic Cloud is reproduced by courtesy of the South African Astronomical Observatory.

▌ Forgotten Constellations

Most people can recognize at least some of the constellations. From Britain, for instance, Orion and the Great Bear are familiar features of the night sky, while from Australia or New Zealand there is the Southern Cross. But how many people have thought about the names of the various groups? Few of them bear any resemblance to the objects after which they are named. It woud take a very lively imagination to picture Orion as a hunter or Ursa Major as a bear, while the Southern Cross is more like a kite.

The first thing to remember is that the stars in any constellation are not necessarily associated with each other, because the stars are at very different distances from us. Look, for example, at Ursa Major. The two end stars of the Bear's 'tail' are Mizar and Alkaid. They appear to lie side by side in the sky, but Mizar is only a little more than half as remote as Alkaid, so that we are dealing with nothing more significant than a line-of-sight effect. Another case is shown by the two brightest stars in Orion, the orange-red Betelgeux and the brilliant white Rigel. Betelgeux is about 520 light-years from us, Rigel about 900. (One light-year is the distance travelled by a ray of light in one year; it is equivalent to nearly six million million miles.) Even more significantly, there are the two Pointers to the Southern Cross, Alpha and Beta Centauri, which are never visible from Britain because they are too far south in the sky. Alpha, which shines as the third brightest of all the stars, is a mere 4.3 light-years from us; Beta is very remote, at well over 400 light-years — yet to the casual observer it seems as though they are near-neighbours.

The constellation patterns thus mean nothing at all, which is another reason for dismissing the pseudo-science of astrology. We can in fact make up what patterns we like. We happen to follow the system worked out by the Greeks more than two thousand years ago. If we had chosen, say, the Chinese or Egyptian patterns our sky-maps would look completely different, though the actual star-positions would be the same. For example, the Egyptian constellations did not include Orion or a Great Bear, but recognized a Hawk and a Hippopotamus!

The last great astronomer of Classical times, Ptolemy of Alexandria, died about A.D. 180. He left us a star catalogue, and also a magnificent textbook which has come down to us by way of its Arab translation; we call it the *Almagest*. Ptolemy listed 48 constellations, most of which

9

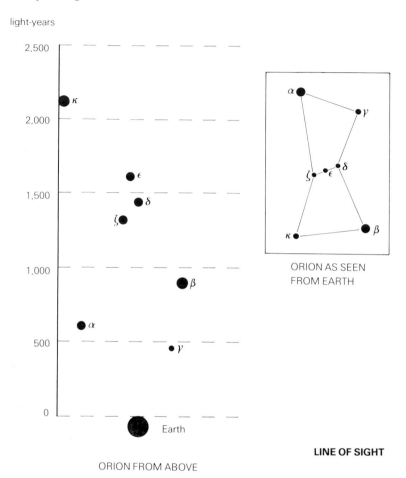

light-years

ORION AS SEEN FROM EARTH

LINE OF SIGHT

ORION FROM ABOVE

were given mythological names. Many of the Olympian gods and heroes are commemorated in the sky, and the legends are fascinating.

One of the most famous stories is that of Perseus and Andromeda. It is said that Queen Cassiopeia, wife of King Cepheus, boasted that her daughter Andromeda was more beautiful than the sea-nymphs. As the nymphs were the daughters of the ocean god (Neptune), this was clearly tactless, and Neptune sent a monster to ravage the kingdom. Consulting the Oracle, Cepheus learned that the only way to appease the god was to chain Andromeda to a rock by the sea-shore, so that she could be devoured by the monster. This was duly done; but at the eleventh hour the situation was saved by Perseus, who had been on an expedition to kill the Gorgon, Medusa, who had snakes instead of hair, and whose mere glance could turn any creature into stone. Perseus was flying home, using winged sandals which had been kindly loaned to him by a friendly Olympian, when he saw the chained princess. Instantly he swooped down, turned the monster to stone by showing it

ORION Note Betelgeux to the upper left, Rigel to the lower right, and the Great Nebula below the Hunter's Belt.

Medusa's head, and then, in the best storybook tradition, married Andromeda. All the main characters in the legend are to be found in the sky – even the sea-monster, Cetus.

Orion too has his legend. He was a great hunter, who boasted that he could kill any creature on earth; but he had forgotten the humble scorpion, which crawled out of a hole in the ground, stung him in the heel, and caused his untimely demise. When both were elevated to celestial rank they were prudently placed on opposite sides of the sky, so that there could be no further unpleasantness. However, not all Ptolemy's 48 constellations were mythological; for instance, we have a Triangle and an Altar.

Ptolemy lived in Alexandria, which is to the north of the Earth's equator. This meant that he could not see the stars of the far south, which never rose above his horizon; when the first maps of these were made, new constellations had to be introduced. Johann Bayer, who drew up a star-catalogue in 1603, was responsible for several, including the 'Southern Birds' — the Crane, the Peacock, the Toucan and the Phoenix. Among other additions were some with decidedly modern names, such as the Telescope and the Microscope, though these came rather later (remember, we have no proof that telescopes were made before 1608, though they may well have been). All these names are to be found on modern maps, though in general the Latin versions are used; thus the Crane is 'Grus', the Microscope is 'Microscopium', and so on.

This was all very well; but as time went by various astronomers introduced more and more groups, some of them with curious names. Johann Elert Bode, in the 1780s, was responsible for such constellations as Globus Aerostaticus (the Balloon), Officina Typographica (the Printing Press) and Lochium Funis (the Log Line). In many cases too fresh constellations were formed by 'stealing' stars from older groups. One of these was Crux Australis, the Southern Cross, which had previously been included in Centaurus, the Centaur. Admittedly Crux is nothing like a cross, but it is at least distinctive, whereas other proposed additions, such as Lilium (the Lily) had no bright stars and no definite shape at all.

Politicians and monarchs were not forgotten. Bode added groups such as Sceptrum Brandenburgicum (the Sceptre of the House of Brandenburg) and Honores Frederici (the Honours of Frederick, otherwise Friedrich II of Prussia), while Edmond Halley, of comet fame, created Robur Carolinum or Charles' Oak, to commemorate the tree in which the later King Charles II hid after his defeat by Cromwell's Roundheads at the battle of Worcester in 1651.

Sky-maps began to look decidedly confused. Things were made

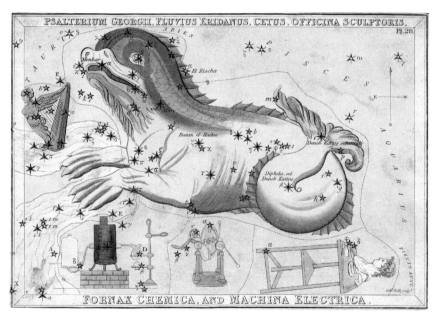

FORGOTTEN CONSTELLATIONS – AND SOME WHICH SURVIVE

worse by constellation-formers such as a monk, Julius Schiller, and
Wilhelm Schickard of Tübingen. Schiller wanted to re-name the
twelve Zodiacal constellations after the twelve Apostles, while Schickard converted all the Classical groups into Biblical ones — so that
Perseus became David, carrying Goliath's head instead of Medusa's,

13

while Hercules was turned into Samson, and Corvus (the Crow) into Noah.

By now things were out of control. Also some of the old constellations were inconveniently large, notably Argo Navis, the Ship Argo in which Jason and his companions set out in their rather unprincipled quest of the Golden Fleece. Subsequently Argo was chopped up into a Keel, Sails and a Poop. At last, in 1932, the International Astronomical Union, the controlling body of world astronomy, lost patience, and revised the whole system. Ptolemy's original 48 constellations were retained, though with altered boundaries, and another forty were also retained, so that the total number of accepted constellations was reduced to 88. Gone were Sceptrum Brandenburgicum, Honores Frederici, Robur Carolinum and the rest.

It cannot be said that the system is satisfactory even now, because the constellations are so unequal in size and importance, but at least it is manageable, and it is not likely to be altered again. Meanwhile there is one rejected constellation which has not been completely forgotten. This is Quadrans, the Quadrant, which had been formed by faint stars in the neighbourhood of Ursa Major (the Great Bear) and Boötes (the Herdsman). Quadrans itself was one of the casualties; but each year, around 3 January, a shower of meteors is seen to radiate from the position in the sky where Quadrans used to be, and we still refer to these meteors as the Quadrantids. So Quadrans lives on, even though the actual constellation has passed into history.

---

2 The Colours of the Stars

'Twinkle, twinkle, little star,
How I wonder what you are . . . . .'

So runs the old rhyme. Today we know what the stars are; they are suns, and our own Sun, which looks so magnificent in our sky, is nothing more than an ordinary star, far smaller and less luminous than many of the stars visible with the naked eye on a clear night. But to most people the stars look alike, apart from obvious differences in brightness. At a casual glance, almost all of them seem to be white.

Of course, stars 'twinkle' or scintillate. This has nothing to do with the stars themselves; twinkling is due entirely to the Earth's unsteady atmosphere, which so to speak 'shakes' the light around as it passes through the air. Stars which are high above the horizon twinkle much less than those which are low down. Sirius, the brightest star in the sky (not counting the Sun!), is the supreme 'twinkler', mainly because of its exceptional brilliance but also because as seen from Britain it is never very high up. On the other hand, I saw it quite recently from South Africa, when it was almost overhead. It still twinkled.

Over the years I have had many letters from people who tell me that Sirius is changing colour. It does admittedly seem to flash various hues when it is very low, but actually it is a pure white star. It is, incidentally, one of our closest stellar neighbours, and is no more than 8.6 light-years away from us; of the really conspicuous stars, only the far-southern Alpha Centauri is closer.

Planets twinkle less than stars, because they show up as small disks instead of points of light, but they are still not immune. In general the planets are easy to identify, but Saturn — and Mars when at its faintest — can easily confuse the unwary observer.

Look more carefully at the stars, and you will see that some of them are definitely coloured. Thus Betelgeux, the bright star in the upper left of the Orion pattern*, is orange-red; so is Antares in the Scorpion, while Capella in Auriga (the Charioteer) is yellow, like the Sun, and Vega in Lyra (the Lyre) is steely blue. It is also easy to see that Dubhe, the brighter of the two Pointers to the Pole Star, is somewhat orange, whereas Merak, the fainter Pointer, is white. On the other hand, it is also true that definite hues are noticeable with the naked eye only for bright stars. Binoculars or telescopes bring out the colours much more clearly. Look, for instance, at the relatively dim Mu Cephei, not far from the north pole of the sky. With any optical aid, its redness is striking; it has been likened to a glowing coal, and the great observer Sir William Herschel nicknamed it 'the Garnet Star'.

The reason why the colours of the fainter stars do not show up with the naked eye is simply because the light-level is too low. For a down-to-earth demonstration of this sort of effect, consider the appearance of two cars, one yellow and one red, as seen by moonlight. You will not be able to tell which is which, because the moonlight is not strong enough (it would take around half a million full moons to equal the light of the Sun). As soon as optical aid is introduced, there are many vividly-coloured stars within range. Of these, the orange and red hues are the most striking.

The colour differences are due to real differences in surface temperature. Red heat is less intense than yellow heat, which in turn is

*From the Northern Hemisphere, that is to say. From Southern Hemisphere countries Betelgeux is seen at the lower right of the Orion pattern.

THE ORION NEBULA This is a photograph taken by John Mason and the author with 1,000 ASA film on a 400 telephoto lens, guided by Patrick Moore's 15–in. reflector.

less intense than white or blue; thus Vega is hotter than Capella or the Sun, while Betelgeux and Antares are cooler. To make up for this, some of the red stars are of immense size. Betelgeux is big enough to contain the whole path of the Earth round the Sun; its diameter is over 250,000,000 miles.

It used to be thought that red giant stars such as Betelgeux were young, and would gradually shrink, heating up to become smaller white or blue stars. Today this is known to be wrong. The red giants are not youthful; they have used up their main reserves of energy, and have reached old age. In perhaps 5,000 million years from now, our own Sun will turn into a red giant. As it will then send out at least a hundred times as much energy as it does at present, the Earth cannot hope to survive. Fortunately, all this lies in the remote future; there is no immediate cause for alarm!

Photographs of nebulae and galaxies taken with very large telescopes show glorious colours. Probably the best pictures of this kind ever taken have been obtained by David Malin, using the 153-inch Anglo-Australian Telescope at Siding Spring in New South Wales; reds, greens, blues and yellows make superb spectacles. Unfortunately, the low level of illumination means that these colours are not directly visible to an observer at the eye-end of a telescope, so that people looking at (say) the Orion Nebula are apt to be disappointed. Yet the colours are real enough, and even photographs taken with a modest telescope, of the sizes used by many amateurs, can bring them out quite well.

Planets, of course, show obvious colour; thus Jupiter and Saturn are yellow as seen through any telescope, while Uranus is green, and Mars has the strong red hue which led to its being named in honour of the God of War. And if you want to see the varied colours of the stars all you need do is take a pair of binoculars, arm yourself with a star-map, and go out to explore the sky on any dark night. There is plenty of variety; it is quite wrong to believe, as so many people do, that one star looks exactly like another.

3 The Legacy of IRAS

In January 1983 a particularly important man-made moon was launched. It was known as IRAS, the Infra-Red Astronomical Satellite, and it operated for the best part of a year before its transmitting equipment failed. It sent back information of all kinds, and all in all it was amazingly successful.*

Infra-red is what we normally term 'heat'. (If you switch on an electric fire, you will feel the infra-red well before the bars become hot enough to glow.) It had long been known that there were many infra-red sources in the sky, but IRAS carried out a full survey, and discovered a quarter of a million more. They were of various kinds, some relatively near at hand and others so far away that their light, or their infra-red radiation, takes thousands of millions of years to reach us.

In our own Solar System, IRAS detected five new comets. One comet which was already known — Tempel 2 — was also studied in detail, with interesting results.

The comet was originally found by the German astronomer Ernst Tempel in 1873. It has a period of $5\frac{1}{3}$ years, and has been seen regularly for well over a century now, though it is always faint. It was not known to have a tail; but when the IRAS results were studied,

---

*Some of these results were given in Chapters 35 and 36 of *The Sky at Night 8* (1985), but it may help to summarize them here. Also, I have updated this chapter from the original television broadcast, because since then there have been important developments — particularly in connection with brown dwarf stars.

IRAS An artist's impression of the Infra-red Astronomical Satellite, far above the Earth, surveying the sky for sources of infra-red radiation.

researchers at Leicester University found a long 'dust' tail, streaming away from the head, which was quite invisible optically but could be picked out in infra-red. Other comets did not seem to have tails of this sort, and it may be that Tempel 2 is exceptional, though it would be hasty to come to any final conclusions.

IRAS also found two belts of thinly-spread material in the Solar System, at about the distance of the asteroid belt (that is to say, between the orbits of Mars and Jupiter). These belts are inclined to the main plane of the Solar System by a full ten degrees, so that they lie to either side of the main plane by a distance equal to that between the Earth and the Sun. It has been suggested that the material may have been produced by a collision between an asteroid and a comet, but again we cannot yet be sure.

Yet it was in studies of bodies well beyond the Solar System that IRAS really came into its own. In particular, it may just possibly have helped us to answer the age-old question: Do other stars have planetary systems like ours?

In the late summer of 1983 two American members of the IRAS team, Drs Hartmut ('George') Aumann and Fred Gillett, were at the

Rutherford–Appleton Laboratory in Oxfordshire, where data from the satellite was being received. They were using various stars as sources for calibrating the infra-red equipment when Dr Gillett suddenly said, 'Hey, Alpha Lyrae has a huge excess!' Alpha Lyrae — Vega — is one of the brightest stars in the sky, 26 light-years away and more than 50 times as luminous as the Sun; it has a hot surface, and is decidedly blue in colour. But it is much younger than the Sun, so that it would not be expected to be associated with cool material which would be detectable in infra-red. At first, the experimenters thought that there must be a mistake somewhere; but when other stars showed nothing similar, they were forced to the conclusion that the infra-red excess round Vega was real.

It was suggested that the material might be due to mass flowing out from Vega, but this was soon ruled out. The material was not moving away from the star, and clearly it had been there for a long time. Further studies showed that it was in particle form, with some of the 'dust' particles much larger than those usually found in interstellar space. The infra-red radiation came from a region stretching out to more than 7,000 million miles from Vega, so that there had to be a considerable quantity of it. 'If there are small particles round Vega,' Dr Aumann told me, 'there must be large particles also. Very careful analyses have been made of what are called fragmentation products. When I applied these methods to Vega, I was astonished because it seemed that the total mass was much the same as that of all the planets in our Solar System combined.' The temperature was estimated as −184 degrees C, which is about the same as that of the icy rings of Saturn.

Were we, then, finding planet-forming material, or even a completed system of planets? Aumann and Gillett quickly came to the conclusion that very small particles would long since have been drawn back into the star, leaving intermediate and large-scale débris in orbit. It was all most intriguing.

The next step was to examine other stars to see if similar infra-red excesses could be found. Fomalhaut, in Piscis Australis (the Southern Fish), gave positive results; like Vega, it is a hot white star, this time about 13 times as powerful as the Sun, and considerably closer than Vega (22 light-years from us). Before long, over twenty cases were established. Of these, probably the most significant was that of Beta Pictoris.

Beta Pictoris is in the southern hemisphere of the sky. It can never be seen from Britain, because it stays below the horizon; it lies close to the brilliant Canopus, which shines as the brightest star in the sky apart from Sirius, and is very luminous and remote. Beta Pictoris itself, in the obscure little constellation of the Painter, is only of magnitude 3.8, so that it is fainter than the dimmest of the seven stars making up the familiar Plough, and it has not been honoured with a proper name. It too is white, with a surface hotter than that of the Sun; its estimated distance is 78 light-years, and it is about 50 times as luminous as the Sun, comparable with Vega. Outwardly there is nothing special about it, and it had always been regarded as a very run-of-the-mill star.

However, the large infra-red excess prompted two American astronomers, Bradford Smith and Richard Terrile, to examine the star at optical wavelengths. For this purpose they went to the Las Campanas Observatory in Chile, which is equipped with a fine 100-inch reflecting telescope. With the 100-inch, they used what is termed a CCD or Charge-Coupled Device — a piece of electronic gadgetry which is far more sensitive than any photographic plate. When they examined Beta Pictoris the results were staggering. They found that there is a circumstellar disk of material extending out to 48,000 million miles from the star. The material is thought to be composed of ices, carbonaceous substances and silicates, which are the very materials making up the Earth and the other planets of our Solar System.

As with some of the other stars examined from IRAS, it was found that the Beta Pictoris cloud of material showed a 'depleted region' extending from 20 to 30 astronomical units from the star itself. (One astronomical unit is the distance between the Earth and the Sun: approximately 93,000,000 miles.) Dr Dana Backman, of the Kitt Peak Observatory in Arizona, commented that this rarefied part of the cloud could be produced by the presence of planet-sized objects which would 'sweep up' much of the dust.

With Beta Pictoris, the cloud is almost edgewise-on to us, and so looks like a fairly thin line; it may be no more than a few hundred million years old. By comparing the amount of heat radiated by the cloud to the amounts which would be expected from particles of different sizes at various distances from the star, a relatively depleted region of the central dust cloud can be inferred. To quote Dr Backman: 'In the case of Beta Pictoris, Earth-sized or larger objects are required. My impulse is that objects in the interior of the cloud are likelier to be planet-sized rather than something bigger, like a dwarf star.' The inner particles, close to the star, may already have been swept away by orbiting planets which exist there.

If all this is correct, what are the chances of life? It is tempting to speculate, but we must be cautious. Beta Pictoris is more energetic than the Sun; if there are planets they are presumably younger than those in our Solar System, so that advanced life-forms may not have had enough time to develop. On the other hand, there is nothing at all

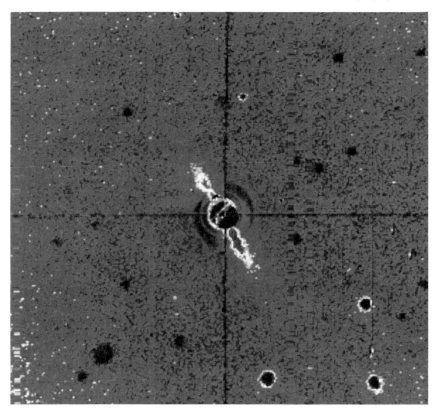

THE OPTICAL VIEW OF BETA PICTORIS

exceptional about Beta Pictoris, so far as we can tell, and evidence is growing that planetary systems are likely to be common in the Galaxy — presumably in other galaxies also.

This leads us on to another problem: what is the essential difference between a large planet and a low-mass star, and are there any 'missing links'? Infra-red researches can help here too, though most of them have been carried out by ground-based observers rather than from IRAS.

Consider a star such as the Sun (or, for that matter, Beta Pictoris). It begins by condensing out of an interstellar cloud of dust and gas known as a nebula. As it shrinks, under the influence of gravitation, its core heats up. When the central temperature has reached the critical value of about 10,000,000 degrees C-nuclear reactions begin. Hydrogen makes up a large percentage of the star's material; when the core has become hot enough the nuclei of hydrogen atoms begin to band together to form nuclei of helium. It takes four hydrogen nuclei to make one nucleus of helium, and each time this happens a little energy

21

is set free and a little mass is lost. The Sun is losing mass at the rate of 4,000,000 tons per second, though there is plenty left; as we have seen, it will be at least 5,000 million years before the Sun starts to run short of available hydrogen 'fuel'.

With a star which is much less massive than the Sun, the central temperature will never become high enough for nuclear reactions to be triggered off. Therefore the star will simply shine feebly because it is contracting, and eventually it will fade out, turning into a cold, dead globe. This is the type of star known — rather misleadingly — as a brown dwarf.

Brown dwarfs have long been known to be theoretical possibilities, but obviously they are very hard to detect even if they are close to us on the cosmic scale, because they are so cool.

Then in 1984, Donald McCarthy and Frank Low, working at the Kitt Peak Observatory in Arizona, announced that they had found a true brown dwarf — a companion of the red dwarf VB8 (VB standing for the Dutch astronomer Van Biesbroeck, who first drew attention to the star).

VB8 lies in the constellation of Ophiuchus, the Serpent-bearer. It is 21 light-years away, and compared with Beta Pictoris, or for that matter with the Sun, it is very feeble indeed. McCarthy and Low claimed to have detected a companion (which they called VB8B) which was smaller than the planet Jupiter, with a diameter of no more than 80,000 miles, but with around 30 times Jupiter's mass; its distance from VB8 itself was given as approximately 600 million miles.

What exactly could it be? It did not seem massive enough to be a normal star, but from its brightness it seemed too 'heavy' to be a planet, so that a brown-dwarf state was the obvious answer. The surface temperature would be of the order of 2,000 degrees.

McCarthy and Low assumed that VB8B was moving round its parent star. Obviously it was difficult to see, and they had to use a modern technique called speckle interferometry, which involves taking a series of very short-exposure images and then sorting them out electronically so as to get rid of the blurring caused by the Earth's unsteady atmosphere. Everything seemed to be in order. Of course, it was impossible to determine the mass of VB8B directly by simply obtaining images of it, but there seemed no obvious reason for any major error.

Later, serious doubts arose. Using a sensitive infra-red detector together with the powerful telescope on the summit of Mauna Kea, in Hawaii, a search was carried out by three more American astronomers: F. Skrutskie, William Forrest and Mark Shure. (Mauna Kea is an ideal site for ground-based infra-red work; I will have more to say about it later.) If VB8B were as cool as expected, it should radiate in infra-red, but nothing could be found. Neither did it show up when looked for by French observers from Chile. VB8B seemed to have disappeared as effectively as the hunter of the Snark, and McCarthy, one of the original co-discoverers, admitted that 'it wasn't where it ought to be'.

What was the answer? Speckle interferometry is a highly delicate technique, and could produce misleading images. Alternatively, what had been seen was not a star at all, but a cloud of gas or small particles associated with VB itself. Thirdly, there was a chance that VB8B was larger and faster-moving than had been thought, and had lined up with its parent, so that it would be temporarily out of view.

There matters rested until 1988, when Benjamin Zuckerman and Eric Becklin, using infra-red techniques from Mauna Kea, produced what may be more reliable evidence for a genuine brown dwarf. It is associated with the white dwarf star Giclas 29–38, which is 46 light-years away. The brown dwarf is around 125,000 miles in diameter. This is less than twice the size of a planet such as Saturn, but it is still larger than the parent star around which it revolves. The brown dwarf is fairly dense — you could pack several tons of its material into an eggcup — but not nearly so compact as the white dwarf, which is a very old star near the end of its luminous career. (Our Sun will become a white dwarf eventually.)

Certainly the object seems to be exceptional, and it could turn out to be the 'missing link' which we have been seeking. The surface temperature is probably no more than 800 degrees C. I would not suggest going for a walk there, but at least it is cool by stellar standards.

I am afraid this has been very much of a digression, so let us now turn back to 1983 and the results obtained direct from IRAS.

In infra-red, the view of the sky is very different from that at optical wavelengths. If our eyes were sensitive to radiation at a wavelength of 12 microns (that is to say, twelve-thousandths of a millimetre), the brightest objects would be old, cool stars together with stars which are surrounded by shells or cocoons of 'dust', the infra-red in this case being due to the dust-shells themselves.

There is a tremendous amount of dust between the stars, and though this blocks out visible light it is no barrier to infra-red. This means that IRAS was able to provide a good picture of the shape of the Galaxy, which has a marked central bulge. At a wavelength of 100 microns the scene is dominated by regions in which star-formation is going on, and there are also stars which are surrounded by what are termed reflection nebulae, composed of dust together with gas. One of these stars is

known to astronomers as Alpha Camelopardalis. It lies in the obscure
constellation of the Giraffe, not far from the Great Bear; to us it looks
unremarkable, so that an effort of the imagination is needed to
remember that it is at least 30,000 times as powerful as the Sun. Its
distance is almost 3,000 light-years.

There were also unexpected developments in connection with what
has become known as 'interstellar cirrus', again due to dust scattered
between the stars. IRAS found that this material was at a strangely
high temperature. This would be expected in the immediate vicinity of
a star, but not in 'deep space', and the reason is not yet known with
any certainty. It may be that the floating dust-grains are smaller than
has been believed, so that they are more easily heated — just as equal
amounts of energy will heat a cupful of water to a higher temperature
than with a whole bathtub.

IRAS reported infra-red sources in areas where absolutely nothing
could be seen visually. More intensive studies have now shown that
these apparently blank areas are almost always occupied by faint
galaxies, which are very remote and highly luminous — millions of
millions of times more powerful than the Sun — and show signs of
being violently disturbed. One of these galaxies is Arp 220. It lies at a
distance of about 300,000,000 light-years, and may well be a system in
which intense star-formation is in progress, so that we are witnessing
what may partly be called a 'starburst'.

Arp 220 was known before the flight of IRAS, and was then thought
to be made up of two galaxies in collision. It now seems that this is
wrong, and that what we are seeing is a single system crossed by an
obscuring dust-lane. It is one of a pair of galaxies, and there may well
be tidal interaction between the two. If so, there could be a resemblance to a much closer and better-known galaxy, Messier 82 in the
Great Bear, which is irregular in shape and is less than 9,000,000 light-
years away. It is the smaller companion of Messier 81, which is just
visible with good binoculars as a faint blur. Apparently material from
Messier 81 is streaming down on to the nucleus of Messier 82, so that
the shock-waves produced by the compression trigger off energetic
star-formation there. But though Messier 82 is a strong infra-red and
radio emitter, it is nothing like so powerful as Arp 220.

We have to admit that our knowledge of the processes going on
inside these systems is very far from complete, and there is a strong
possibility that some of them at least may be powered by super-massive
central black holes, produced by old, collapsed stars which are now
pulling so strongly that not even light can escape from them — though
material which is just about to be sucked down into the 'forbidden
area' is so violently heated that it gives off a tremendous amount of

radiation. So long as we have to depend upon visual observations we are hopelessly handicapped, which is why satellites such as IRAS are so important.

IRAS itself came to the end of its career in late 1983, and it no longer transmits, though it is still circling the Earth and will continue to do so for many years to come. Within the next decade it will be succeeded by ISO, the Infra-red Space Observatory, which is being planned by the European Space Agency. No doubt we will be treated to many new surprises.

We have come a long way since 1957, when the Russians launched the first of all artificial satellites: Sputnik 1, which was no larger than a football and which carried little apart from the radio transmitter which sent out the never-to-be forgotten 'Bleep! bleep!' signals. As each year passes we learn more about the Solar System, the stars and their evolution, the interstellar 'dust' and the remote galaxies. Astronomy has made more progress during the past thirty years than it did during the previous thirty centuries.

4 The Rills of the Moon

Look at the Moon, even with the naked eye, and you will see the dark patches which we still miscall 'seas' even though we know that there has never been any water in them, and that they are simply plains of volcanic lava. Binoculars bring out the craters — walled formations which dominate the entire lunar scene — together with mountain ranges, isolated peaks, domes and valleys. Because the Moon is so much smaller and less massive than the Earth, its gravitational pull is much weaker, and it has been unable to hold on to any atmosphere it may once have had. The Moon today is an airless world, so that there are no winds, no rain, no clouds and no weather.

The most impressive mountain ranges form parts of the borders of the regular 'seas' (in Latin, *maria*). Thus the Lunar Apennines make up part of the boundary of the vast Mare Imbrium, or Sea of Showers. There are also many isolated mountains and clumps of peaks, but the main emphasis is always on the craters, which range from tiny pits too small to be seen from Earth up to vast enclosures more than 150 miles in diameter. Some of them have central mountains and terraced walls; others have smoother floors, and in many cases the originally circular forms have been broken and distorted. Indeed, we can make out what

ORBITER PICTURE OF THE HADLEY RILL ON THE MOON

are termed 'ghost craters' which have been almost obliterated by later surface activity.

The craters are everywhere. They fill the bright uplands, and they are also to be found on the dark Mare regions. From Earth we can see only part of the total lunar surface; the Moon spins on its axis in exactly the same time that it takes to complete one orbit (27.3 days), so that the same hemisphere faces us all the time. To be more precise, there are slight 'wobblings' or librations which allow us to examine a grand total of 59 per cent of the surface, but the remaining 41 per cent remained unknown until 1959, when the Russians sent their space craft Luna 3 on a round trip and obtained the first pictures of the averted

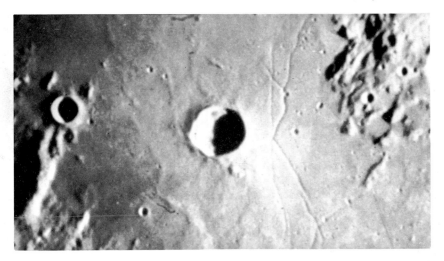

THE TRIESNECKER CLEFTS ON THE MOON

regions. Since then we have been able to draw up detailed maps of the whole of the Moon, and we know that the far side is just as mountainous, just as barren, and just as crater-scarred as the side we have always known. (Incidentally, the equality of the orbital period and the axial rotation period is not coincidence. Tidal friction over the ages has been responsible, and we know that the major satellites of other planets have similarly synchronous or 'captured' rotations.)

Among the minor features of the Moon are the valley-like clefts, rills or rilles. Some of them, such as Schröter's Valley near the brilliant crater Aristarchus, are easy to see with small telescopes when well placed, and it is clear that not all are of the same type. Many of the so-called rills prove to be crater-chains, made up of small craters which have run together, often with the destruction of their common walls, giving the impression of strings of beads. We also have the linear rills, and the very significant sinuous rills.

The linear rills are usually fairly regular and straightsided. They must have been produced by tensional stresses in the Moon's crust which resulted in fracturing, so that the ground between two parallel fractures subsided. The sinuous rills are different. Superficially, they look rather like old river-beds, but they were not cut by water; analyses of the lunar samples brought home by the Apollo astronauts and the Russian unmanned probes have told us that there has never been any water on the Moon. However, the sinuous rills were undoubtedly cut by some liquid, and the only possibility seems to be volcanic lava.

The Earth today is a volcanic planet; the sight of a major eruption —

27

such as happens often enough in Hawaii — is breathtaking. Molten magma wells up, and cascades down in streams of fiery lava. The Moon's active history, on the other hand, ended long ago, but there can be no doubt that intense vulcanism took place in the distant past. The lava-streams heated the ground and melted it, so that eventually the result was a channel — in fact, a sinuous rill. If the lava-flow were less turbulent, the end-product would be a visible lava-flow, and these are indeed common on the lunar seas.

The Moon today is virtually inert, and any lingering activity is on a very small scale indeed. So far as we can tell, the main lava-flows took place between 3,000 and 4,000 million years ago, so that by geological standards even a 'young' flow is very ancient indeed. It is instructive to compare the situation with that on Mars, which is of course larger and more massive than the Moon, and cooled down much more slowly. On the Martian surface we see lofty volcanoes, one of which, Olympus Mons, rises to a height of some fifteen miles above the adjacent landscape; there are also channels which seem genuinely to have been cut by running water. But there are also sinuous rills which look very like those of the Moon, and it seems that these must have been formed in the same manner.

By contrast, the Earth has cooled down more gradually still, and remains very hot inside. No sinuous rills are being formed now, because eruptions do not continue for long enough, but there is some evidence that they may have been so formed in the remote past. Venus, about the same size as the Earth, has been found to have craters and valleys; we cannot see them directly, because of the planet's dense, cloud-laden atmosphere, but they have been mapped by the radar equipment carried in Russian and American probes. However, no lunar-type sinuous rills have been identified on Venus, and it may well be that they do not exist there.

Even though men have been to the Moon, it would be wrong to claim that we have anything like a complete knowledge of the lunar world. There is still disagreement about the origin of the main craters; many astronomers believe that they were formed by meteoritic bombardment, while others prefer to think that they were produced by internal action. For that matter, we are also unsure of the origin of the Moon itself, though the idea that it once formed part of the Earth has fallen into disfavour. It is more than fifteen years since the end of the Apollo programme, and we cannot yet say when the next lunar trips will take place, though they are hardly likely to be delayed for more than a decade at the most, and the idea of a fully fledged Lunar Base by the end of the century is not in the least far-fetched.

Of one thing we can be certain: there has never been any life on the

Moon. Its age is known to be about the same as that of the Earth (around 4,600 million years), and it has always been sterile. Long before the dinosaurs appeared on our own world, the Moon had become quiet, and the great volcanoes had been silenced. Yet we can still see the results of the tremendous activity which must once have occurred there, and the winding sinuous rills are of special significance.

5 Mission to Amphitrite

There can be nobody who has not heard of the bright planets — Mars, Venus and the rest. But how many people know the name of Amphitrite? Probably very few, apart from astronomers, because although Amphitrite is a member of the Solar System it is not a proper planet. It is one of the swarm of minor planets or asteroids, and by no means the largest of them, but during 1985 it leaped into prominence. It was scheduled to be the first asteroid to be surveyed from close range by a space-craft.

Any casual glance at a map of the Solar System shows that it is divided into two parts. First there are the four small planets Mercury, Venus, the Earth and Mars, all of which are solid and rocky. Beyond the orbit of Mars there is a wide gap, beyond which come the four giants Jupiter, Saturn, Uranus and Neptune (plus Pluto, a peculiar body which does not seem to fit happily into the general pattern). In the 1770s Johann Elert Bode, of Germany, publicized a mathematical relationship linking the distances of the planets from the Sun which is still known as Bode's Law, though Bode himself was not the first to draw attention to it, and in any case it is probably due to nothing more than coincidence. However, Bode's Law was taken very seriously at the time, and it indicated that there ought to be another planet moving in the gap between the paths of Mars and Jupiter. In 1800 a group of astronomers calling themselves the 'Celestial Police' set out to find the missing planet. Between 1801 and 1807 they found not one, but four small planets moving almost in the middle of the gap; they were named Ceres, Pallas, Juno and Vesta. Ceres, the first of them, was in fact discovered by G. Piazzi, from the Palermo Observatory in Sicily, before the 'Police' had started full-scale operations — hence its name, in honour of the patron goddess of Sicily.

All these asteroids, as they came to be called, were very insignificant

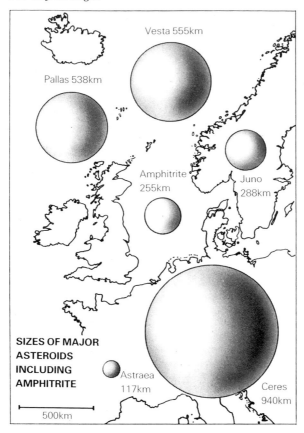

Vesta 555km

Pallas 538km

Amphitrite 255km

Juno 288km

SIZES OF MAJOR ASTEROIDS INCLUDING AMPHITRITE

Astraea 117km

Ceres 940km

500km

judged by planetary standards. Ceres, much the largest of them, is less than 700 miles in diameter, and only Vesta is ever visible with the naked eye. No more small worlds seemed to be forthcoming, and eventually the 'Police' disbanded. Then, in 1845, a German amateur named Hencke discovered Asteroid No.5, Astraea; in 1847 two more came to light, and after that year the discoveries came at a furious rate. By now more than three thousand asteroids have had their orbits worked out, though not many of them are as much as fifty miles in diameter, and the smallest members of the swarm must be mere lumps of material only a few hundred feet across. Even if combined, the known asteroids would not make up one body as massive as the Moon.

How were they formed? It used to be thought that they represented the débris of an old planet (or planets) which met with disaster, by collision or explosion, but this intriguing theory has now been more or less abandoned. The main planets were produced from a 'solar nebula', a cloud of material surrounding the youthful Sun, but no large planet could form in what is now the asteroid zone, because of the disruptive gravitational pull of mighty Jupiter.

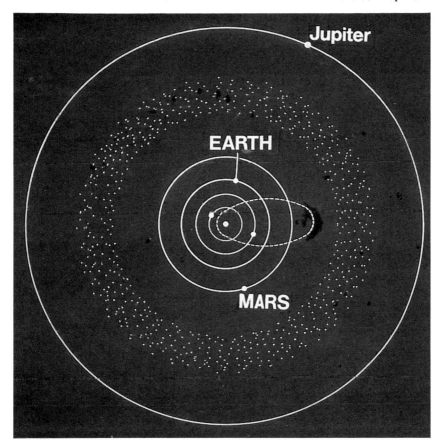

THE ASTEROID ZONE, BETWEEN THE PATHS OF MARS AND JUPITER

There are some asteroids which swing away from the main swarm, and may approach the Earth; in 1937 the mile-wide midget Hermes brushed past us at a mere 485,000 miles, about twice the distance of the Moon, and it is not inconceivable that at some stage we could be hit by an asteroid. Indeed, this has almost certainly happened in the past. Two asteroids, Icarus and Phaethon, approach the Sun to within the orbit of Mercury, while in the outer part of the Solar System there is one asteroidal oddity, Chiron, which spends most of its time between the orbits of Saturn and Uranus.

Now let us consider Amphitrite, which has a diameter of 158 miles and is therefore one of the major members of the swarm. A modest telescope will show it as a star-like point, though it is much too faint to be seen with the naked eye (it seldom rises above the tenth magnitude). It is in the midst of the main zone, and moves round the Sun at a distance ranging between 220,000,000 miles and 253,000,000 miles; recall that the mean distance of Mars from the Sun is 141,500,000

31

A PHOTOGRAPH OF AMPHITRITE BY R. W. ARBOUR; THE ASTEROID IS ARROWED

miles, and that of Jupiter 483,000,000 miles. The revolution period of Amphitrite is four years. It is a darkish body, reflecting only about 13 per cent of the sunlight which falls upon it.

There seemed nothing to single Amphitrite out from its fellows — except that it happened to be in a position suitable for contact by a space-probe. In May 1986 the Americans were scheduled to launch their Galileo mission towards Jupiter, the main object of which was to survey Jupiter and its satellites. Obviously Galileo had to pass through the asteroid zone, and on 6 December 1986, after a journey of nearly 350,000,000 miles, it would have been within range of Amphitrite. As the asteroid was passed, at a distance of no more than 6,000 miles, Galileo's cameras would have been able to send back high-resolution pictures of it. The chance to survey an asteroid was clearly not to be missed.

Alas for such hopes! The disaster with the *Challenger* space-shuttle, in January 1986, caused chaos in the whole United States space programme. One of the casualties was Galileo. By now (1989) it has been rescheduled, but it will not reach Jupiter until late 1995, and it will not go anywhere near Amphitrite, though a couple of smaller asteroids may be surveyed. For the moment, therefore, we can do no more than speculate as to what Amphitrite is really like.

First, there is not the slightest chance that it or any other asteroid can hold on to even a trace of atmosphere. The gravitational pulls are far too weak, and if you could stand on the surface of Amphitrite you would weigh less than this copy of the book you are reading. Obviously, then, life of any sort is out of the question. Yet we are very anxious to know what the surfaces of the asteroids are like, and what materials exist there. It is very likely that there will be craters — just as there are on the two dwarf satellites of Mars, Phobos and Deimos, which are quite possibly ex-asteroids which were captured by Mars in the remote past — but more than that we cannot say at the moment.

From Amphitrite, the sky will be inky black. The constellation patterns will be the same as those we see from Earth, but there will be additions to the overall view, because many asteroids will be visible with the naked eye. Mars, of course, will be an inferior planet, and will show phases just as Venus and Mercury do to us, but any Amphitritan astronomer would have a poor view; the distance between Mars and Amphitrite is always over 65,000,000 miles. The Earth, from over 125,000,000 miles, will be inconspicuous, and will seem to keep very close to the Sun.

It cannot be said that the asteroids have always been popular members of the Solar System. Photographic plates exposed for quite different reasons are often found to be crowded with asteroid images, all of which have to be eliminated and which waste an amazing amount of time. One irritated German observer went so far as to describe them as 'vermin of the skies' (which I regard as most unkind, particularly as one small asteroid, No. 2602, discovered by Dr Edward Bowell from Flagstaff Observatory some years ago, has been officially named 'Moore' after me!). But small though they may be, the asteroids are of real interest, and we are very anxious to learn more about them. It is a pity that progress has been so badly held up, but before too long we should have close-range information from some of the members of the asteroid swarm, and one day we should even be able to draw a map of the surface of Amphitrite.

6 Lift-off from Kourou

On 2 July 1985 I was privileged to witness a particularly important event in the history of space exploration. From Kourou, in French Guyana, the Giotto probe to Halley's Comet lifted off to begin a journey which would reach its climax in March of the following year.

Halley's Comet, our most famous celestial visitor, comes back for our inspection every 76 years. After its return in 1910 it vanished into the farther reaches of the Solar System, but in 1982 it was recovered as a tiny speck in the sky, and by 1985 it was well on its way back to the Sun. Everybody heard about it, but not everyone had any real idea of a comet's nature.

The commonest mistake is to suppose that a comet streaks quickly across the sky, travelling from horizon to horizon in a few seconds. In fact, all comets are very distant — far beyond the top of the Earth's atmosphere — and they do not move perceptibly against the starry background. If you see an object shifting obviously, it cannot be a comet. It will either be an artificial satellite, of which thousands have been launched since the Space Age opened with the ascent of Russia's Sputnik 1 in October 1957, or else a meteor (unless, of course, it is something much more mundane, such as a weather balloon or a high-flying aircraft).

Comets are bona-fide members of the Solar System, but they are quite unlike planets. They are not solid and rocky; a large comet consists of an icy central nucleus, a head or coma, and a tail or tails made up of tiny particles of 'dust', together with extremely thin gas. Though comets may be of immense size — the head of the Great Comet of 1843 was larger than the Sun — they are very flimsy, since the only reasonably massive part of the comet (its nucleus) is no more than a few miles in diameter. Even a direct collision between the Earth and a comet could do no more than local damage.

Comets move round the Sun, but in almost all cases their orbits are very elliptical, and with one exception (Halley's) all the really bright comets take hundreds, thousands or even millions of years to complete one circuit. This means that we cannot predict them, and they are always liable to take us by surprise. During the last century we were greeted by several brilliant comets, such as those of 1811, 1843, 1858, 1861 and 1882, but in our own time they have been depressingly rare, and the last really 'great' comet was that of 1910, though of course

KOUROU BASE,
FRENCH GUYANA
Giotto was launched
from here towards
Halley's Comet by an
Ariane rocket.

there have been many others which have been conspicuous enough to be seen with the naked eye.

There are also many short-period comets with periods of only a few years. Encke's Comet, named in honour of the German astronomer who first calculated its orbit, goes round the Sun in only 3.3 years. But these short-period comets are faint, and usually remain well below naked-eye visibility. Moreover, they generally lack tails, and appear as nothing more than tiny, fuzzy patches.

Halley's Comet is in a class of its own. We know that it has a period of 76 years, and it has been seen regularly since well before the time of Christ; there is even a Chinese record of it dating back to 1059 BC. Yet the regularity of its appearances was not known until the work of Edmond Halley, Britain's second Astronomer Royal, and it is only right that the comet should be named in his honour.

Halley was born in London in 1656. He came of a well-to-do family, so that he was sent to St Paul's School (where he became captain) and went on to Oxford to take his degree. There was never any doubt about his ability, and he was also one of those rare people who is almost universally liked. He had a jovial disposition, and was an amusing companion; he was entirely free of malice or jealousy, in which he differed sharply from some of his contemporaries. There are many anecdotes about him. For instance, he certainly knew the Tsar of

Russia, Peter the Great, who came to England in 1698 to learn about shipbuilding. It is said that after a far from teetotal evening the Tsar climbed into a wheelbarrow and Halley pushed him through a hedge. Whether the story is true or not, there is no doubt that Halley would have been perfectly capable of such a thing — and we do have his receipt for two damaged wheelbarrows . . . . At one stage in his career he made several ocean voyages, mainly to study the Earth's magnetism, and one of his few enemies — John Flamsteed, at that time Astronomer Royal — made the jaundiced comment that 'Halley now talks, swears, and drinks brandy like a sea-captain', which he probably did.

Halley first became known in the scientific world when he went to the island of St Helena to make observations of the southern stars, which never rise over Europe. Then, in 1682, he observed a bright comet. Nobody then knew just how comets moved, or even what they were; it was usually thought that they travelled in straight lines, visiting the Sun only once before moving out into the depths of space. Halley was not so sure. A few years later Isaac Newton published his immortal *Principia*, in which he laid down the laws of gravitation (incidentally, it was Halley who persuaded Newton to write the book, and also paid for the publication out of his own pocket). Later still, in 1705, Halley used the new methods to work out the orbit of the comet of 1682. He found that it followed almost precisely the same path as those of comets previously seen in 1607 and in 1531. Could the three

HALLEY'S COMET IN THE BAYEUX TAPESTRY The comet blazes down, while King Harold topples on his throne!

36

comets be one and the same? Halley believed so. If the revolution period were 76 years, the comet would be seen again in 1758. He wrote: 'If it should return again in 1758, posterity will not refuse to acknowledge that this was first discovered by an Englishman.'

Halley died in 1742, but on Christmas Night, 1758, the comet was recovered in the expected position in the sky by a German amateur named Palitzsch. It passed perihelion — its closest point to the Sun — in 1759, since when it has been seen again in 1835, 1910 and the present time.

Once the period had been worked out, it became possible to check old records and identify earlier returns. (Note that the period is not absolutely constant, and may vary by a year or two either way — because a body as flimsy as a comet is easily perturbed by the gravitational pulls of the planets, particularly giant Jupiter, and no two orbits are exactly alike.) For example, the comet was seen in 12 BC, though suggestions that it could have been the Star of Bethlehem are completely out of court; the timing is wrong, and in any case the comet could have been seen by anybody and not only the Wise Men. The return of AD 837 was particularly striking, when the comet had a brilliant head and tail stretching more than half-way across the sky. It was seen once more in 1066, some months before Duke William invaded England, and the Saxons regarded it as an evil sign; it is shown in the famous Bayeux Tapestry, with King Harold toppling on his throne and his courtiers looking aghast. At the return of 1456 it caused so much alarm that the current Pope, Calixtus III, preached against it as an agent of the Devil (even though he did not go so far as to excommunicate it, as has been suggested).

Why have comets been regarded as unlucky? The fear of them goes back a long way. Remember Shakespeare, in *Julius Caesar*:

When beggars die, there are no comets seen:
The heavens themselves blaze forth the death of princes.

The fears were partly superstitious and partly practical. It was thought that a head-on collision with a comet might destroy the world, and even as recently as 1910 there were uneasy misgivings. It is quite true that a comet's tail contains some gases which would be poisonous in concentration, but the cometary material is so rarefied that it is harmless, and even when the Earth passed through the tail of Halley's Comet nothing could be detected. This did not prevent some people in Chicago from barricading their doors and windows as a protection against the noxious fumes, while one enterprising salesman made a large sum of money by selling what he called 'anti-comet pills', though exactly what they were meant to do has never been revealed!

HALLEY'S COMET (12 MAY 1910) This was taken from Yerkes Observatory. The comet had a 30-degree tail, and was much more spectacular than in 1986.

Comets shine by reflected sunlight, though when near perihelion it is true that the cometary material does emit a certain amount of light by fluorescence. They can be seen only when moving in the inner part of the Solar System, so that for most of its 76-year period Halley's Comet is too far away to be seen; according to the traffic laws of the Solar System it moves quickest when near the Sun, so that it spends only a few years inside the orbit of Mars. After the 1910 return it was lost until October 1982, when it was picked up again by astronomers at Palomar, in California, using the great 200-inch reflecting telescope there. It was extremely faint, but brightened steadily as it moved inward towards perihelion on 9 February 1986.

Unfortunately, it as clear from the outset that this would be a most unfavourable return — the worst for over two thousand years — so that the comet would never become a bright naked-eye object, as it had been in 1835 and in 1910. However, space-probes could be sent to it, and no less than five were planned years ahead of time. The Americans abandoned their mission because of the cost (something they will undoubtedly regret for the next seventy-six years), but there were two Japanese probes, two Russian and one European. It was a veritable space armada. For once there was no competition, and the whole programme was a splendid example of international co-operation.

The Russian probes, Vega–1 and Vega–2, were launched in December 1984, and went first past the planet Venus; in June 1985 they dropped balloons into the atmosphere of that decidedly hostile world before continuing their journey to the comet. Vega–1 was scheduled to make its closest approach to Halley on 6 March 1986, leaving Vega–2 to follow three days later. Their main tasks were to investigate the conditions in the immediate vicinity of the comet, and to find out just where the nucleus lay inside the head. The small but highly sophisticated Japanese probes, Sakigake and Suisei, were due to rendezvous at about the same time. Then, on the night of 13–14 March, the European space-craft Giotto — named in honour of the Florentine painter Giotto di Bondone, who saw the comet in 1301 and used it as a model for the Star of Bethlehem in his picture *The Adoration of the Magi* — was scheduled to plunge right into the coma, and study the nucleus from close range. It was an ambitious project by any standards, and there could be no second attempt. Giotto had either to work first time, or not at all.

The rocket-launching base at Cape Canaveral in Florida is well known, and there has also been a good deal of information about the bases in the USSR. Fewer people knew anything about Kourou, or even where it was! It is in fact the second city of French Guyana, with a population of some 8,500; only Cayenne is larger. Guyana itself covers an area of 36,000 square miles, and extends between latitudes 2° and 6° N. Originally it was populated by Indian tribes, and a few of these tribes still live in the forests; French colonies date from the 1670s, and of course became famous (or should one say notorious?) for the penal settlement which included Devil's Island, in the group of isles which can be clearly seen from the Kourou coast. The penal colony came to its unlamented end in 1945. Less than twenty years later, the Space Centre was established. (*En passant*, some of the ex-convicts settled in Guyana after their release, and I am told that one of them now actually runs a restaurant in Kourou.)

As a launching site, Kourou has various advantages. It is so close to the equator that full advantage can be taken of the Earth's rotation, and in the event of any mishap there is little danger of anyone being hurt by a piece of falling débris. The facilities of the CSG — Centre Spatial Guyanais — extend from Kourou itself to Sinnamary, over about thirty miles of the Atlantic coast. Obviously the CSG is a French affair, but it is officially the launching base for all European satellites and probes: it became operational in 1968 with the launch of a sounding rocket (Véronique), and has been used for all the rockets of the Ariane series, beginning in December 1979. It is fair to say that Ariane had become reasonably reliable, though there were one or two failures. The 'Arianespace' organization had reason to be satisfied with its achievements, and it was sensible to entrust the dispatch of Giotto to an Ariane rocket.

By courtesy of British Aerospace — the prime contractors for Giotto; the probe was built in Bristol — I arrived in Guyana on 30 June. The climate seems to range from very hot to blazing hot, and it is also so humid that most people find it unpleasant, but from the space-launching point of view it is good; for lift-off the visibility must be better than 2,000 feet, with clouds not below 800 feet and no danger of lightning strikes. There have been few delays to Ariane rockets on this score.

The base itself is very impressive, and is being extended quickly to prepare for new missions, but on this occasion we were all preoccupied with Giotto. There was only one space-craft, and Halley's Comet would not be within range again for another 76 years, so that everything depended upon Ariane producing a faultless performance. When we arrived preparations were nearing their final stages, and we were assured that there were no problems which could not be solved.

The procedure for the launch, timed for the morning of 2 July, had been worked out to the last detail. After the motors were fired Ariane would lift off in 3.4 seconds. After 2 minutes 30 seconds the first stage, having exhausted its fuel, would break away and fall back into the sea (plans had been made to salvage it, though in the event this was the only part of the programme which did not work). The second stage

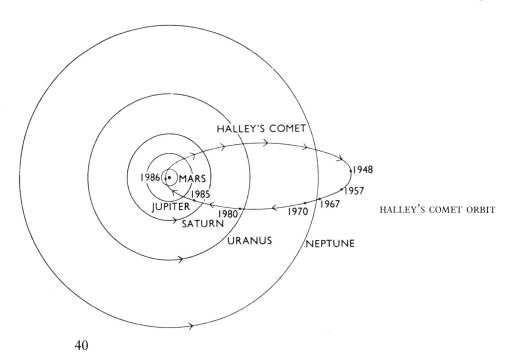

HALLEY'S COMET ORBIT

would separate after 4 minutes 50 seconds, and the space-craft itself would separate from the third stage after 14 minutes 58 seconds. Giotto would then enter what is called a 'geostationary transfer orbit', a closed path round the Earth with a minimum height of 124 miles and a maximum of 22,240 miles. Then after three circuits a small solid-propellant rocket on Giotto would fire, and put the probe into its path towards Halley's Comet. The rendezvous had to take place when the comet's orbit took it across the plane of the orbit of the Earth, so that there could be no deviation from the planned encounter in March 1986. By the time of arrival Giotto would have covered a total distance of about 425,000,000 miles.

My rôle was to broadcast a commentary on the launch itself, and obviously it was essential to watch from a considerable range! Six minutes before the planned lift-off there was a flurry of excitement: a 'hold', and the countdown was stopped. I imagine that this was a particularly bad moment for those who had worked so long and so hard, because nobody knew just what was wrong. Mercifully, it turned out to be nothing more than a faulty pen-recorder on one of the ground stations, and it was put right. The countdown recommenced. Then, at 8.23 am local time, the motors roared. From our viewing station we could see the glow, and then came the sound; Giotto lifted off, and in a remarkably short period had vanished above the clouds which covered most of the sky.

There followed a tense period. If anything went wrong now, nothing

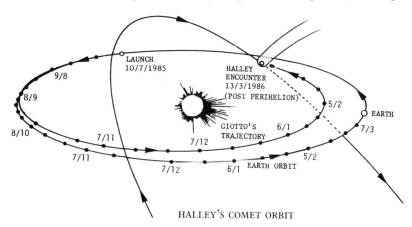

HALLEY'S COMET ORBIT

Reference trajectory for Giotto from launch on 10 July 1985 to post-perihelion encounter on 13 March 1986.

Halley's orbit is inclined at 162°; it crosses the ecliptic twice the first time (from south to north) on 9 November 1985, the second time on 11 March 1986.

Five phases of Giotto Mission

1. Launch and geostationary orbit
2. From geostationary orbit to first mid-course manoeuvre
3. Cruise
4. Pre-encounter (5 days)
5. Encounter (4 hours)

could be done. Fortunately, all was well. The first and second stages separated on schedule; then came the separation of Giotto itself — and a spontaneous burst of clapping from all those in the control-room, where I was, and in the outer viewing positions. Ariane had completed its task, and Giotto was in the correct orbit. Final confirmation came on the following day, when the small 'booster' also fired faultlessly, increasing the probe's speed by 0.93 miles per second and breaking it free from the Earth for ever.

Giotto was on its way. I wonder what Edmond Halley would have said?

7 Jack and his Rider

Of all the constellations in the northern hemisphere of the sky, the most famous is undoubtedly Ursa Major, the Great Bear. Its seven chief stars make up the pattern always known as the Plough (in America, the Dipper), and over England they never set, so that they are always on view whenever the sky is sufficiently dark and clear.

The stars are not outstandingly bright, but the pattern makes them distinctive. Two of them, Dubhe and Merak, are known as the Pointers, because they show the way to the Pole Star in the Little Bear, which lies within one degree of the polar point. The others are Alkaid, Mizar, Alioth, Megrez and Phad. Megrez appears much the faintest of the seven; Alkaid is the most remote and the most luminous.

Five of the Plough stars are moving through space in the same direction at the same rate, forming what is termed a moving cluster. Alkaid and Dubhe are moving in the opposite direction, so that eventually the Plough will lose its shape — though the individual or 'proper' motions are so slight that no change in outline is noticeable over a period of many human lifetimes.

The most interesting of the Plough stars is Mizar, otherwise known as Zeta Ursae Majoris. It is the second star in the 'handle' and it is easily identified, because it has a much fainter star, Alcor, close beside it. Alcor is an easy naked-eye object under even reasonable conditions; it and Mizar have often been nicknamed 'Jack and his Rider'.

Mizar and Alcor are genuinely associated, but there are some differences of opinion about their distances. One recent catalogue makes them 20 light-years apart, which is well over 100 million million miles. Other estimates make them about the same distance from us, around 88 light-years, in which case they are less than one light-year apart. Certainly they share a common motion through space.

There is a minor mystery associated with the pair. The Arab astronomers of a thousand years ago were very keen-sighted, and yet they regarded Alcor as a severe test of vision. There is nothing difficult about it today; it can be seen without the slightest trouble, even by people who do not have acute eyes. So what is the answer?

One rather unexpected suggestion has been made. Between Mizar and Alcor is another star, of magnitude 8 and therefore quite beyond naked-eye range, though a small telescope will show it. It was named 'Sidus Ludovicianum' in 1723 by admirers of the Landgrave Ludwig V, under the impression that it was a newly born star or even a new planet! Can it be that Sidus Ludovicianum is variable, and that in ancient times it used to be brighter than it is now? Frankly, I doubt it. It is a perfectly ordinary star, lying in the background and not connected with the Mizar–Alcor pair, but we still cannot tell why the Arabs regarded Alcor as so elusive.

In 1651 the Italian astronomer Riccioli made an important discovery. Using one of the low-power telescopes common at the time, he looked at Mizar, and found that the bright star itself was made up of two components, so close together that with the naked eye they appear as one. This was the first telescopic double star. The magnitudes are 2.3 and 3.9, and the separation is $14\frac{1}{2}$ seconds of arc, not much too close to be split with good modern binoculars.

Many thousands of double stars are now known, but not all are genuine pairs. In the so-called optical doubles, one component lies almost behind the other, so that there is no true association. Physically-associated pairs are termed binaries, in which the two members are revolving round their common centre of gravity (much as the two bells of a dumbbell will do if you twist them by the joining bar). Rather surprisingly, binaries are much commoner than optical doubles. Mizar is one such system; the two components are always at least 35,000 million miles apart, which is 380 times the distance between the Earth and the Sun. The revolution period must amount to thousands of years. The brighter component, Mizar A, is 70 times as luminous as the Sun.

Riccioli's discovery proved to be only the start. In 1889 E.C. Pickering, in America, found that Mizar A is itself a binary. He could not see the components separately, because they were much too close, but he could track them down by means of the spectroscope.

Just as a telescope collects light, so a spectroscope splits it up. Pass a beam of sunlight through a glass prism, and you will find that it is spread out into a rainbow, with red at the long-wave end and violet at

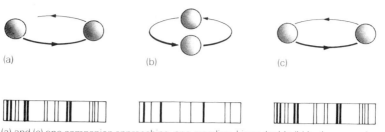

(a) and (c) one companion approaching, one receding. Lines double (b) both companions moving across line of sight. Lines single

THEORY OF SPECTROSCOPIC BINARY STAR

the short-wave end. It is the same with the stars, which are themselves suns. The rainbow band is crossed by dark lines, each of which is the particular trademark of one special substance; thus two dark lines in the yellow part of the rainbow must be due to sodium, and to nothing else. By using the spectroscope, we can tell which chemical elements are present in the stars. (In 1830 the French philosopher Auguste Comte stated that any knowledge of the chemistry of the stars would be for ever beyond Man's capabilities — which only goes to prove that when a French philosopher makes a profound statement, he is almost certain to be wrong!)

This is not all. If an object is moving away from us, all the lines in the rainbow will be shifted over to the long-wave or red end; if the object is approaching, the shift of the lines will be towards the blue and violet. This is the well-known Doppler effect. The shifting of the lines in the spectrum of Mizar gave Pickering the clue. The lines regularly became double; then single; then double once more.

If Mizar is a close binary, with the two components moving round their 'balancing point', there will be times when their movement in orbit is transverse to our line of sight; there will be no Doppler shifts, and the spectral lines will appear single. But when one component is approaching us the other must be receding; the spectral lines will be shifted — with one component to the red, and with the other component to the blue — so that the lines will become double. Pickering found that this was precisely what was happening. The revolution period turned out to be $20\frac{1}{2}$ days.

Later it was found that the fainter component, Mizar B, is also a spectroscopic binary with a revolution period of 182 days; there may be a third member of the main system — and to complete the picture, Alcor is itself a close spectroscopic binary, with a total luminosity of fifteen Suns.

What would the sky look like from a planet in the Mizar system? Two brilliant double suns, a fainter single one, and a distant double

URSA MAJOR photographed by the author with an ordinary camera. Mizar is the second star from the left; it is easy to see its companion, Alcor.

which would seem like a twin point of light. Whether there can be any planets there is dubious, because Mizar is such a complicated system, but we cannot rule it out. Meanwhile, it is always worth looking at Mizar, with a telescope if you can and with the naked eye or binoculars if not. There is much to be learned from 'Jack and his Rider'.

8   Red Shift — or Red Herring?

Pisces, the Fishes, is not one of the more glamorous constellations of the Zodiac. It adjoins the Square of Pegasus, and consists of a long line of faint stars, none too easy to identify. But there are two faint objects which attracted a great deal of attention in 1985. They were galaxies, known as NGC 7603 and 7603B. (NGC stands for New General Catalogue, though by now it is hardly new; it was drawn up by J.L.E. Dreyer a century ago.)

NGC7603 is the larger of the pair, and like so many galaxies is spiral

45

in shape. 7603B is much smaller, and seems to be joined on to its companion by a luminous bridge made up of gas, dust and stars — something which is by no means unusual. There is a much more prominent example in the Whirlpool Galaxy, one of the closest of the external systems, which has a smaller companion at the end of a similar-looking bridge.

Looking at a photograph of NGC7603 and 7603B there seems no doubt that the two are actual neighbours, lying at the same distance from us. Yet, if we believe the conventional theories, this cannot be true. The red shifts in the spectra indicate that NGC7603 is about 560 million light-years away from us, and is receding at 5,000 miles per second, while NGC7603B is 1,050 million light-years away, with a recessional velocity of 10,000 miles per second. This indicates that the two are not associated with each other, and that the bridge is simply material lying in the same line of sight.

For well over half a century now the concept of the expanding universe has been so widely accepted that to question it has been regarded as nothing short of heretical. It all comes back to the Doppler effect, which as we have seen indicates a movement of approach or recession — a red shift for recession, a blue shift for approach. Even before the First World War, astronomers at the Lowell Observatory in Arizona found that apart from a few of the closest systems (those in what we now call the Local Group), all the galaxies showed red shifts, so that they were racing away — and the farther away they were, the faster they were going. To be more precise, the nature of the galaxies was not then known, but in the 1920s Edwin Hubble, at Mount Wilson, used the 100-inch telescope to outline the picture which has been accepted even since, at least by most people.

With relatively near galaxies there are various ways of measuring distances, but this becomes much more of a problem with galaxies too far off for individual stars to be made out. This is where what we call Hubble's Law comes to the rescue. The red shift in the spectrum tells us how fast the galaxy is receding, and this in turn gives us the distance. There is some disagreement about the precise relationship between velocity and distance — the Hubble Constant — but the usually accepted value is about 55 kilometres per second per mega-parsec,* though some astronomers believe it to be considerably higher. As a rough guide, a galaxy 30 million light-years away will be receding at 500 kilometres per second, while a galaxy 300 million light-years away will be receding at 5,000 kilometres per second. The distances involved are tremendous by any standards. The most remote object so far measured is racing away at 93 per cent of the velocity of light, in

*1 megaparsec = 1 million parsecs; 1 parsec = 3.26 light-years.

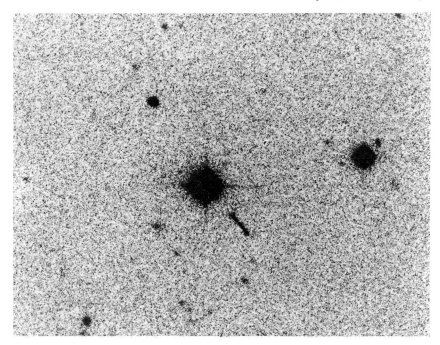

QUASAR 3C-273, DISPLAYING ITS CURIOUS 'JET'

which case its distance from us must be at least 13,000 million light-years.

But . . . . everything depends upon the red shifts, interpreted as pure Doppler effects. If there is some other cause, then our whole theory breaks down, and evidence is growing that this could be true. One great astronomer who believes so is Sir Fred Hoyle. Another is Dr Halton C. Arp, formerly of the Mount Wilson Observatory and now carrying out his researches in Germany. Hoyle, Arp and their supporters believe that the red shifts are misleading, and that many of the outer systems are not so far away as their spectra indicate.

Of special significance are the remarkable objects which we call quasars, first identified as long ago as 1963. Quasars look small, but are incredibly luminous, and all of them show very high red shifts. Conventionally, they are regarded as the centres of very active galaxies, but a single quasar may in this case outshine at least a hundred normal galaxies — and remember that our Galaxy, which is fairly typical, contains about a hundred thousand million stars!

Halton Arp has different ideas. He has found many cases of galaxies and quasars which appear to be lined up, and are presumably associated, but which show very different red shifts. His opponents maintain that this is sheer coincidence, but Arp has also found pairs of galaxies which also seem to be associated and which again show different red shifts. There is even one quasar (Markarian 205) which gives every impression of being embedded in a galaxy, NGC 4318, but has a far greater red shift; and in 1982 Jack Sulentic carried out an analysis which has convinced him that the bridge from the galaxy extends right into the centre of the quasar.

What about our two galaxies in Pisces? Either the bridge is tricking us, or else the red shifts are not pure Doppler effects. Dr Nigel Sharp, at the Kitt Peak Observatory, believes that the bridge can be traced beyond the smaller system, and that NGC7603B really is in the background, but the test is far from conclusive.

There is one possible observational check, involving what is termed velocity dispersion. In a galaxy the component stars are moving at different speeds and in different directions, so that at any one time some of the stars are moving towards us and others are moving away. (Of course, these effects are superimposed upon the overall velocity of the galaxy, which can be allowed for.) The approaching stars have their spectral lines blue-shifted, while the receding stars produce red shifts. As there are thousands of millions of stars in the galaxy, with a wide range of speeds, the lines merge to produce broad lines in the total spectrum. The greater the broadening, the greater the range of speeds, i.e, the velocity dispersion. Measurements of this kind can be used to estimate the galaxy's luminosity, and this in turn gives the distance. The principle is known as the Faber–Jackson relationship.

Sharp applied this test to NGC7603 and 7603B. The result was that the velocity dispersion seemed too low for a galaxy of the power that 7603B ought to be according to its red shift, but unfortunately the uncertainty is too great for any definite conclusions to be drawn. All that can be said is that while Sharp's results are consistent with B being a remote, unconnected galaxy, there is no concrete disproof of Arp's claim of a real association between the two.

Yet there are many other instances of what appear to be anomalous red shifts. One famous case concerns what is known as Stephan's Quintet, a group of galaxies discovered by M.E. Stephan as long ago as 1877. Of the five members, four have similar red shifts, while the fifth is completely different. There is also the 'galaxy chain' VV172 (VV stands for the Russian astronomer Vorontsov-Velyaminov, who drew attention to it). Here there are five systems, two to each side of a larger central member and perfectly aligned; the second galaxy seems, from its spectrum, to be much more remote than the others.

Arp goes so far as to suggest that instead of being the nuclei of active galaxies, quasars may be relatively small objects ejected from galaxies

— and some of them could even be in the Local Group, only a few million light-years from us.

At present the whole question remains open, unthinkable though this would have seemed only a few years ago. Arp and his supporters are still in the minority, but there are serious doubts about the reliability of the red shifts as distance–gauges, and if there really is some other explanation we will have to do some drastic re-thinking. It is not impossible that, after all, the red shifts will turn out to be red herrings.

9 The Zodiacal Light

Comets were very much in the news during 1985 and 1986. Halley's Comet was back, and caused a tremendous amount of interest. We will not see it again until the return of 2061, but at least we will continue to see evidence of it in the form of meteors; two showers, the Orionids and the Eta Aquarids, are associated with it. Meteors are, in fact, cometary débris. But are there any other signs of material left by comets?

The answer is 'yes'. We see them in the form of the Zodiacal Light, a cone which sometimes appears above the western horizon after sunset or in the east before dawn. It is never very prominent from Britain, where the air is seldom transparent and there is an alarming

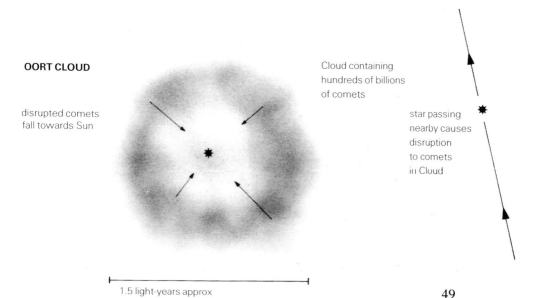

OORT CLOUD

disrupted comets
fall towards Sun

Cloud containing
hundreds of billions
of comets

star passing
nearby causes
disruption
to comets
in Cloud

1.5 light-years approx

amount of light-pollution, but in more favoured climates it can rival the brightness of the Milky Way.

Contrary to popular belief, space is not empty. There is thinly-spread material everywhere, even between the galaxies, and we have abundant evidence of what is termed the 'interplanetary medium' in our own Solar System. Fortunately, it is very tenuous, and is not nearly dense enough to cause appreciable friction; if it were, space-travel would be well-nigh impossible. But it can be measured, and it can also become visible. In ancient times the Zodiacal Light was often referred to as 'the false dawn', and as early as the year 1100 it was described by the Persian poet-astronomer Omar Khayyám, though it was not until 1672 that it was correctly explained — by the Italian G.D. Cassini, best remembered today, perhaps, for his discovery of the main gap in Saturn's ring-system.

The Solar System must be older than the Earth, which has itself an age of around 4.6 thousand million years. It is believed that it was formed out of a cloud of dust and gas, which became flattened as it condensed; the inner part produced the Sun, while the planets and other bodies built by 'accretion' due to gravity. There was a great deal of material left over when the planets had been born. Icy comets were plentiful, and it is thought that there must now be a vast cloud of them orbiting the Sun at a distance of at least six million million miles (one light-year). There was also general débris, and this thinly-spread material was concentrated near the main plane of the system. Remember that the planets move in more or less the same plane; apart from Pluto which may have no real claim to planetary status, none of the planetary orbits have inclinations of more than 8° to that of the Earth, and only two (Mercury and Venus) have inclinations of more than 3°. If you draw a plan of the Solar System on a flat piece of paper, you are not very far wrong.

It is the sunlight illuminating this interplanetary 'dust' which produces the Zodiacal Light. Strictly speaking, it forms a dimly luminous band, but its brightness falls away quickly with apparent increasing angular distance from the Sun, so that it gives the impression of a cone rising from the place where the Sun has set or is about to rise. This means that it is a phenomenon of either the late evening or the early morning. Obviously, it extends along the Zodiac: hence its name.

The particles making up the Zodiacal Light are very small indeed, with diameters of only around two micrometres; the low density is demonstrated by the fact that according to the best estimates there is only about one particle of this size in a cube of space measuring more than 300 feet along each side. The total mass is very slight, and cannot be more than 1/10,000,000,000 of the total mass of the planets.

One interesting point is that the particles cannot remain in orbit permanently. If they are above a certain diameter they are driven out by the pressure of the Sun's radiation, while those which are not so expelled gradually spiral inward to destruction because of what is termed the Poynting–Robertson effect. Interactions between the particles and solar radiation makes them gradually lose momentum, so that eventually they are drawn into the Sun and are destroyed. In view of this, how can we explain the presence of any interplanetary medium at all?

Clearly it must be constantly enriched in some manner, and the answer seems to lie with the comets, which leave material in their wake as they travel round the Sun. The larger particles may be seen as meteors when they dash into the upper part of the Earth's air and are burned away by friction; most of the annual meteor showers are known to have their parent comets (Halley's Comet, as we have noted, is responsible for two). The reliable Perseids of August are the débris left by a less brilliant comet, Swift–Tuttle, which has a period of well over a century and has not actually been seen since 1862, while the April Lyrids are associated with Thatcher's Comet of 1861. (The Thatcher who discovered it was an American astronomer, and his comet will not return for over 400 years, so that there are no political affiliations whatsoever!) The smaller particles, too small and too lightweight to produce luminous effects when they enter the atmosphere, are widely

COMET HALLEY, PHOTOGRAPHED FROM DURBAN, 13 MARCH 1986

diffused along the plane of the Solar System, and it is these which are responsible for the Zodiacal Light.

Comets come in regularly from the distant Oort Cloud, and several new ones are discovered each year. Unless they are suitably perturbed by a planet (usually Jupiter), or unless they fall into the Sun (as sometimes happens; several cosmic suicides have been recorded, from space-probe photographs), they will return to the Cloud, not to be back for many centuries, or even thousands or millions of years; but as they near the Sun they become active, and shed 'dust' in large quantities. The result is that the supply of Zodiacal particles is continually replenished. There is one theory, due to Victor Clube and Bill Napier, that the Sun collects a new supply of comets each time it passes through a spiral arm of the Galaxy, and that there is a subsequent period when comets are exceptionally plentiful, so that the Zodiacal Light is more prominent too; they even suggest that this may have been the case in near-historical times, so that our remote ancestors may have seen really brilliant displays. This idea is not widely supported by astronomers in general, but it is certainly intriguing.

Obviously, the Zodiacal Light is at its best when the ecliptic is most steeply inclined to the horizon at the appropriate time. This is always the case in the tropics, where the celestial equator crosses the sky at high altitude and so, therefore, does the ecliptic; but things are less favourable from higher latitudes. In Britain, for example, the best times are evenings in February or March, and mornings in September and October. Unfortunately, it is essential to keep well clear of artificial illuminations. Anyone who lives in a city, or even in a village with obtrusive street lights, has little hope.

Even more elusive is the Zodiacal Band, which extends farther along the ecliptic and is really an extension of the Light. In 1854 the Danish astronomer Theodor Brorsen described a patch of light which can be seen in the sky exactly opposite to the Sun, and which may be quite large even though it is extremely faint; it is usually known as the Gegenschein (or, in English, the Counterglow), and it is of the same nature as the Zodiacal Light. (Brorsen was not the first to see it, since it had been reported much earlier by Humboldt, but it was Brorsen who gave the first truly scientific account of it.) The Gegenschein is by no means easy to find. From England I have seen it only once; that was in 1942, when the whole of the country was blacked out as a precaution against German air-raids, and the sky was pleasingly dark and clear.

Dim though they are, the Zodiacal Light and the Gegenschein are of real importance. They are worth searching for, and to glimpse them is always a source of personal satisfaction.

10 How wrong we were!

Astronomy today is a fast-moving science. We have learned a great deal during the past few years, and new techniques and new methods have led to progress which would have seemed out of the question half a century ago. But there have been mistakes too; and now and then it is useful to look back and see where astronomers — or, at least, some of them — went wrong.

The Moon provides one good example of what I mean. In the 1950s Dr Thomas Gold, one of the world's most celebrated astronomers, put forward a strange theory according to which the so-called lunar 'seas' were filled with soft dust; he even wrote that a space-craft unwise enough to land there would promptly sink out of sight with devastating permanence. Practical lunar observers (such as myself) were sceptical, because the theory did not fit the facts as we saw them, but it was taken very seriously in America, and it was not finally disproved until the first unmanned space-ships and then human astronauts landed there.

Venus was more of a problem. It is about the same size as the Earth, and is closer to the Sun than we are, so that it must be expected to be hotter; but as its surface is permanently hidden by a layer of cloud,

THE MOON FROM LUNA 9 This was the first successful soft-lander, which came down in 1966 in the Oceanus Procellarum. It finally disposed of the 'dust' theory.

Earth-based observations did not tell us much. Before the Space Age it was usually thought that the 'day' there was about as long as a terrestrial month, whereas it is actually equal to 243 days, longer than Venus' own 'year'. There was also a theory, due to F.L. Whipple and D.H. Menzel, that the planet was covered mainly with water, with only a few isolated islands here and there.

This would have led to a bizarre situation. The atmosphere of Venus contains large amounts of the heavy gas carbon dioxide, and so any surface water would have been fouled, producing oceans of soda-water. However, it was recalled that in the early stages of the Earth's existence our air had much more carbon dioxide in it than it has now, and it was during this period that life began — in the seas. Could the same sort of process be operating on Venus at the present time?

It was an attractive idea, but the various space-craft, beginning with Mariner 2 in 1962, showed that it was completely out of court. The surface temperature of Venus is not far below 1,000 degrees Fahrenheit, so that no liquid water can possibly exist. Venus is a bone-dry dust-desert, with active volcanoes and almost constant thunder and lightning, so that it is very like the conventional picture of hell. Sir Fred Hoyle had suggested that there might be oceans of oil; these too had to be ruled out.

Just why Venus is in this state is not known with real certainty, but it must be due essentially to the lesser distance from the Sun: a mere 67,000,000 miles, as against 93,000,000 miles for the Earth. According to one theory, the Sun used to be much less powerful than it is now, so that Venus and the Earth started to evolve along similar lines. Then, slowly but inexorably, the Sun became hotter. Earth was moving at a safe distance; Venus was not. The oceans there evaporated, the carbonates were driven out of the surface rocks, and in a relatively short time Venus became the raging inferno of today; any life which had appeared there was ruthlessly snuffed out. Whether this is correct or not, we can hardly hope to know until it becomes possible to analyze samples of the surface materials. I very much doubt whether we will find any fossils, but I may be wrong.

There was a major error, too, concerning the innermost planet, Mercury, which is 36,000,000 miles from the Sun on average, and has a revolution period of 88 days. The two greatest planetary observers of pre-war days, G.V. Schiaparelli and E.M. Antoniadi, believed that the rotation was synchronous — that is to say, the same as the revolution period — in which case Mercury would always keep the same face turned towards the Sun, just as our Moon does with respect to the Earth. There would be an area of permanent 'day' and an area of permanent 'night', with only a narrow intermediate zone over which

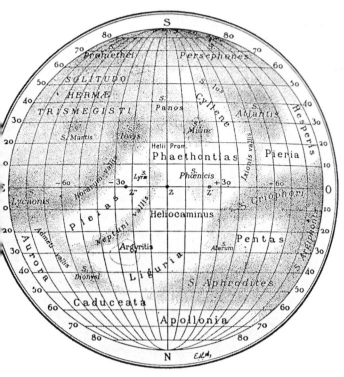

ANTONIADI'S MAP
OF MERCURY
Through no fault
of his, the map
bears little
relation to the
truth, and his
nomenclature has
been abandoned.

the Sun would rise and set, bobbing up and down over the horizon. Science-fiction writers made great play of the Mercurian 'twilight zone', stressing that while the sunlit region was torrid the dark area would be colder than any other planet in the Solar System.

Fascinating — but, again, wrong. Infra-red observations and radar studies showed that the true rotation period is 58.6 Earth days, or two-thirds of a Mercurian year, so that all parts of the surface are in daylight at some time or other. Yet Schiaparelli and Antoniadi cannot be blamed. By a strange locking effect, Mercury turns the same hemisphere towards us every time it is best placed for observation, so that the vague surface markings detectable with powerful telescopes always appear to be in the same positions on the disk. When Mercury is less well placed the surface features cannot be seen at all. In any case, the results from the later probe Mariner 10, which made three active passes of the planet in 1973 and 1974, showed that the surface is cratered like that of the Moon, while the older maps were so inaccurate that they had to be abandoned completely.

Mars, with its red deserts, its dark features and its white polar caps, is a very different kind of world. Around 1960 I recall giving a lecture at London University during which I made a series of statements about Mars, every one of which was backed up by the best available scientific evidence, and every one of which turned out to be incorrect! It was thought that the dark areas were old sea-beds filled with organic matter ('vegetation', if you like); that the surface was no more than gently undulating, with no major valleys or mountain chains; that the polar caps were very thin deposits of hoar-frost, only a millimetre or two thick; and that the Martian atmosphere was made up chiefly of nitrogen, with a ground pressure of about 85 millibars — much the same as the pressure of the Earth's air at a height of 52,000 feet above sea-level.

In fact, the Mariner and Viking probes have shown that far from being smooth, Mars is heavily cratered, with giant volcanoes and deep valleys. One volcano, Olympus Mons, rises to a full fifteen miles above the outer landscape. There was no trace of the famous Martian canals, and the dark areas were not depressions; the most prominent of them, Syrtis Major, is a lofty plateau. The darkness is due simply to the scouring away of the red 'dust' by the Martian winds, so that the surface below is exposed. The atmosphere is made up principally of carbon dioxide, but the ground pressure is below 10 millibars everywhere, so that conditions approximate to what we usually regard as a vacuum.

What about life? The Vikings scooped in material from the surface, analyzed it, and sent back the results. It must be said, with regret, that no positive signs of biological activity were found, and it is starting to

A TRENCH BEING DUG ON MARS BY VIKING I The reddish surface was analyzed in a search for life, but no positive signs were found.

look as though Mars is sterile, at least at the present time. This disposed of a persuasive argument by the Estonian astronomer Ernst Öpik, who had claimed that the dark regions had to be due to something which lived and grew, as otherwise the whole surface would soon be covered with the reddish material from the deserts.

Another casualty was the alleged 'wave of darkening' during Martian spring and early summer, which had been reported by many experienced observers. It was maintained that when the polar cap started to shrink, with the onset of warmer weather at the end of winter, the dark regions near the poles began to show signs of activity, becoming more prominent and more well defined; the process went on until even the regions near the equator had been affected. I had always been openly sceptical about the wave of darkening, because I had never been able to see it myself, but it would have made sense if the dark patches had really been due to vegetation which started to develop when the moisture from the polar cap reached it. Since there is no vegetation, and no wave of darkening, I am rather glad that I failed to record it!

There was one rather amusing 'aside'. In the late 1950s the famous Rusian astronomer Iosif Shklovsky had suggested that the two dwarf satellites of Mars, Phobos and Deimos, were not natural bodies, but were artificial space-stations built by the Martians. He also claimed that life was likely to be widespread throughout the universe, both inside and outside the Solar System. Astronomers in general were not impressed with the Phobos and Deimos theory, and the Mariners pictures showed both satellites to be irregularly shaped rocky masses, pitted with craters. Much later, in 1982, Shklovsky wrote an article for the *Yearbook of Astronomy*, at my invitation, and told me that the Phobos story had been nothing more than a practical joke, while he now believed that there was no life anywhere except on Earth. Well . . .

Next on the list were the giant planets Jupiter and Saturn, each of which produced its quota of surprises. First Pioneers 10 and 11, then Voyagers 1 and 2, by-passed Jupiter and sent back information from close range, much of which was unexpected.

Pioneer 10 showed the way, and made its closest approach to Jupiter in December 1973. It was found that the planet's magnetic field was amazingly strong, and that Jupiter was associated with belts of radiation so powerful that the Pioneer equipment was nearly saturated and put out of action; the paths of all the later probes were suitably modified, to avoid the worst of the danger.

The famous Great Red Spot came under careful scrutiny. It had been thought to be either a solid body floating in Jupiter's atmosphere, or else the top of a column of stagnant gas. The Pioneers and Voyagers showed otherwise; the Red Spot is a whirling storm — a phenomenon of Jovian meteorology — and the red hue seems to be due to phosphorus.

Then there were the four large satellites, Io, Europa, Ganymede and Callisto, known collectively as the Galileans because they were studied by Galileo, with his primitive telescope, in early 1610. All are easy telescopic objects; Io is slightly larger than our Moon, Europa slightly smaller, and Ganymede and Callisto much larger — in fact, Ganymede is actually larger than the planet Mercury, though less dense and less massive.

Rather naturally, it had been assumed that the Galileans must be very like each other, with icy, probably cratered surfaces. Nothing could have been further from the truth. Certainly Ganymede and Callisto fit this description, but Europa's surface is a smooth ice-sheet, with shallow cracks which criss-cross the whole satellite and make it look rather like a hard-boiled egg which has been dropped on to the floor. Io is even more remarkable. It is red and sulphur-covered, with violently active sulphur volcanoes which are erupting all the time, and are extremely hot. Since it also moves inside Jupiter's radiation zone, it must be just about the most lethal world known to us.

Jupiter was found to have a thin, dark ring, but of course it is Saturn which is always known as the ringed planet, and I always regard it as the loveliest object in the whole of the sky. There were known to be three main rings, two bright (A and B), and one dusky and semi-transparent (C). Rings A and B are separated by the Cassini Division, and in Ring A there is a narrower division, Encke's. Other reported divisions had been officially discounted. Gerard Kuiper, one of the pioneers of planetary exploration, had examined Saturn with the Palomar 200-inch reflector, and had said positively that no genuine gaps in the rings existed apart from the Cassini Division; even Encke's was dismissed as a mere 'ripple'.

In November 1980, when Voyager 1 made its rendezvous with Saturn, I was at the Jet Propulsion Laboratory in Pasadena, California, which is the headquarters of all NASA planetary missions. As soon as the pictures started to come in I shared in the general amazement. The rings turned out to be incredibly complex; far from being smooth sheets of icy particles, they were made up of thousands of narrower ringlets and gaps. Not all the rings were perfectly circular, and Ring B showed strange radial 'spokes' which defied explanation.

My instinctive feeling was that we were seeing some sort of wave effect produced by Saturn's satellites in the ice-sheets, and this does seem to be correct, though the full details remain to be worked out.

SATURN'S RINGS, FROM VOYAGER 2 Note the numerous minor divisions, much too delicate to be seen from Earth.

Even the Cassini Division contained narrow rings, whereas we had always believed it to be empty.

Another revelation concerned Titan, which is much the largest of Saturn's satellites. It was known to have an atmosphere, but we had expected this atmosphere to be made up chiefly of methane (marsh-gas), and to be thin. Actually it is thick, with a ground pressure $1\frac{1}{2}$ times that of the Earth's air at sea-level, and the main constituent is nitrogen, together with a good deal of methane. Titan's surface was hidden by the clouds in its atmosphere, and all that Voyager could show us was the top of a layer of what might be termed orange smog, so that we do not yet have any real idea of what conditions on the surface are like. There may be seas of liquid methane, cliffs of solid methane, and a methane rain dripping down all the time from the orange clouds above. At any rate, we may be sure that Titan is unlike any other body in the Solar System, and most of our previous ideas about it were very wide of the mark. Of Saturn's other satellites, one (Mimas) is dominated by a crater about one-third the diameter of the satellite itself; another (Hyperion) is shaped rather like a hamburger, but its longer axis does not point straight towards Saturn, as dynamically it ought to do.

There have been major changes of outlook, too, with regard to the galaxies. For example, the distance of the Andromeda Spiral, the nearest of the really large external systems, had been given as 750,000 light-years, but in 1952 Walter Baade discovered that there had been a major error; the real distance is now known to be 2,200,000 light years. In one short paper, delivered to the Royal Astronomical Society, Baade calmly doubled the size of the entire universe. His paper was greeted with stunned silence; I was there, and I remember it well. Then there were the quasars, which in 1963 were found to be immensely remote and super-luminous (at least, so most astronomers think). They had been recorded long before, but had always been mistaken for normal stars in our own Galaxy.

Please do not misunderstand me. Astronomers have been right much more often than they have been wrong. But it is only natural to make mistakes, and at least it shows that astronomers are human!

I I   The Puzzling Sun

We owe everything to the Sun. It sends us virtually all our light and heat, and without it we could not exist; indeed, the Earth itself would never have been born. Yet as we have seen, the Sun is nothing more than a normal star, shining because of nuclear reactions going on deep inside it; hydrogen is being converted into helium, with the release of energy and loss of mass.

There have been no really major changes in the Sun since the dawn of human history, but in some respects it must be regarded as a variable star. Every eleven years or so it is particularly active, and there are many groups of the dark patches which we call sunspots, together with violent flares and outbreaks. Activity then dies down, until at spot-minimum the disk may be blank for many consecutive days before activity starts to rise once more. The last maximum was that of 1980, so that the next is due around 1991.

The solar cycle is not perfectly regular, and not all maxima are equally energetic. We cannot even be quite sure that the cycle itself is permanently present. According to old records, the period between 1645 and 1715 was characterized by an almost complete lack of spots; this is known generally as the Maunder Minimum, because attention was drawn to it more than eighty years ago by the English astronomer E.W. Maunder.

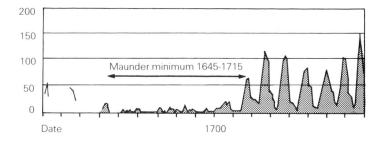

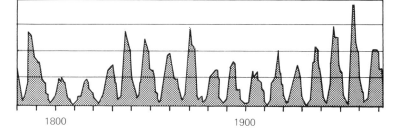

SOLAR CYCLE

It is difficult to be precise, because the records are not complete, but there is plenty of supporting evidence. For example, the state of the Sun affects the form of the corona, the glorious 'pearly mist' which makes up the Sun's outer atmosphere, and is visible with the naked eye only during a total solar eclipse, when for a few minutes the Moon passes right in front of the Sun and blots out the brilliant surface or photosphere. Eclipse reports during the Maunder Minimum indicate that the corona was not nearly so prominent as usual.

There is also a link with aurorae or polar lights, which are due to electrified particles sent out by the Sun which enter the Earth's upper air and cause the lovely displays so familiar to people who live in high latitudes. (Aurorae are common from North Scotland, rather rare from South England and almost unknown from countries close to the equator, though it is claimed, probably correctly, that on one occasion an aurora was seen from Singapore.) When the Sun is active there are greater numbers of emitted particles, making aurorae more frequent. It seems that auroral displays were lacking during the Maunder Minimum; the great astronomer Edward Halley saw his first aurora only in 1715, when the period of inactivity was coming to an end.

It may or may not be significant that the Maunder Minimum coincided with a spell of very cold weather, at least over Europe and probably over the rest of the world as well. During the 1680s the Thames froze almost every winter, and frost fairs were held on it.

61

TYPICAL SUNSPOTS

Going back still further in time, we can trace indications of earlier cold periods coupled with a lack of sunspots, though obviously the records are so fragmentary that it would be unwise to put too much faith in them. There are also suggestions that the Ice Ages which have affected the Earth throughout its history may have been due to slight fluctuations in the Sun's output. The last cold spell ended only 10,000 years ago, and we may even now be in the midst of an 'interglacial', in which case there may well be another Ice Age ahead.

Another problem concerns the so-called solar wind, which is a stream of particles being sent out by the Sun constantly in all directions. It had been expected that the solar wind would slow down with increasing distance from the Sun, but apparently this is not true, as we have found from the data sent back by the Voyager probes to the outer planets. The speed of the solar wind is the same in the remote regions of the Solar System as it is in the neighbourhood of the Earth.

We have to admit that we cannot explain this peculiar state of affairs, and neither do we know the full extent of the heliosphere, i.e., the region inside which the solar wind is detectable. Again we depend mainly upon the Pioneer and Voyager probes, none of which will ever come back, and are on their way out of the Solar System. With luck, we should be able to keep in touch with them until well into the 1990s, perhaps until they reach the edge of the heliosphere; we must hope for the best. Meanwhile, it is worth noting that the solar wind affects the tails of comets, driving gas-particles out of the cometary nucleus and producing straight tails which show tell-tale movements and dislocations when the solar wind is unusually 'gusty'.

Next we must consider neutrinos, which are particles with no mass (or virtually none) and no electrical charge, so that they are very hard to detect. They can be made to interact with atoms of chlorine to produce radioactive argon, and for years now a fascinating experiment has been in progress a mile below ground in the Homestake gold-mine, near Deadwood Gulch in South Dakota. Cleaning fluid contains chlorine, and the experiment is designed to trap the neutrinos which enter the tank of cleaning fluid which has been placed there; it has to be deep down, as otherwise the situation would be confused by the less penetrating particles called cosmic rays, which produce similar effects.

There seems to be no doubt that the Sun is producing far fewer neutrinos than it ought to do. The experiment itself is reliable enough, so that there may well be something wrong with our theories. If the temperature at the Sun's core is 'only' 14,000,000 degrees C, rather less than is generally believed, the neutrino problem can be solved, but this raises a host of other difficulties. There have even been suggestions that the Sun has temporarily 'switched off', which sounds unlikely but cannot be completely ruled out.

If the Sun is not generating as much nuclear energy as we expect, can it be that there is some other power-source — for example, the production of energy by gravitational forces, if the Sun is slowly shrinking? According to accepted theory, the Sun is actually expanding as it grows older, though not by much (the apparent diameter may increase by about 4 seconds of arc in 50,000,000 years). Yet there have been suggestions that only a few centuries ago the Sun was perceptibly larger than it is now. John Eddy and Aram Boornazian, at the United States Naval Observatory, have claimed that its diameter is shrinking at about five feet per hour. If this went on for centuries, it would alter the Sun's output and change the climate of the Earth very much for the worse.

Clearly the shrinkage could be no more than a temporary phenomenon, because we know that the Earth's temperature cannot have altered, overall, by more than about fifteen degrees since the start of life here. (For example, fishes need a relatively stable sea temperature if they are to survive.) But in any case, recent studies have cast serious doubt upon the results by Eddy and Boornazian.

Measuring the apparent diameter of the Sun to the required accuracy is not nearly so easy as might be thought, but there are various methods of attack. One is to use a telescope called a meridian circle, which is mounted on an east-west axis and can swing only in altitude. For many years the position of the Sun in the sky has been regularly measured from national observatories such as Greenwich and

Paris, for navigational purposes. When the Sun crosses the observer's meridian at noon it is timed by the fixed telescope — and, of course, the period taken for the Sun to pass right over the meridian tells its apparent diameter. Another method is to time a transit of Mercury, when that quick-moving little planet passes straight across the face of the Sun. (Transits are not too common; the last was that of 1986, while the next will not be until 6 November 1993.) Finally, we can use solar eclipses. We know the apparent size of the Moon, and at eclipse we can compare it with the apparent size of the Sun.

All in all, it looks as though the alleged shrinkage of the Sun over the past three hundred years is not genuine, and so can give us no help in solving the irritating problem of the missing neutrinos. However, we still have to reckon with recently detected pulsations and 'quiverings' in the Sun, and we have yet to solve the full story of why sunspots behave as they do. We may have to concede that we know much less about the Sun than we fondly believed only a few years ago; there is much that we yet have to learn about our own particular star.

12 Messenger to Uranus

Far beyond Saturn, the outermost member of the Sun's family known in ancient times, there moves the strange green planet Uranus. It was discovered in 1781 by William Herschel; it is just visible with the naked eye if you know where to look for it, but it seems exactly like a star. No Earth-based telescope will show anything definite upon its tiny disk. It takes 84 years to complete one journey round the Sun, moving at an average distance of over 1,780 million miles.

Some basic facts about Uranus had been established before the Voyager 2 mission. It is a giant world, over 30,000 miles in diameter, but it is not in the least like the Earth. It was thought to have a solid core, overlaid by a gas-and-liquid zone containing a great deal of water; above this came the atmosphere, rich in hydrogen and helium. The temperature was believed to be so low that any form of life could safely be ruled out. A system of dark, thin rings had been discovered in 1977, but these were quite different from the glorious icy rings of Saturn, and were extremely difficult to see from Earth; they had been found more or less by accident, when they passed in front of a star and made the star 'wink'. Five satellites were known, all of them smaller than our Moon: Miranda, Ariel, Umbriel, Titania and Oberon.

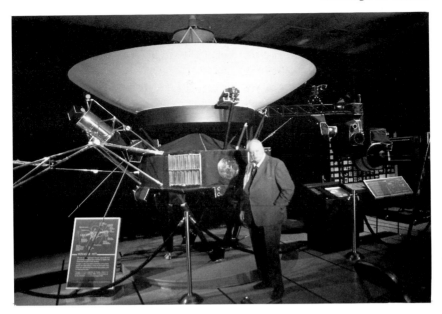

FULL-SCALE MODEL OF VOYAGER 2, AT JPL, WITH THE AUTHOR

Perhaps the most peculiar characteristic of Uranus was its axial tilt. The Earth's axis is tilted to the perpendicular by 23 degrees, which is why we have our seasons; most of the other planets have inclinations of around the same value, though Jupiter's is much less (only 3 degrees) and Venus spins in a wrong-way or retrograde direction for reasons which we do not know. Uranus is different. The axial tilt is 98 degrees — more than a right angle — producing a very remarkable calendar. First the south pole, then the north, has a 'day' lasting for 21 Earth-years, with a corresponding period of night in the opposite polar zone; for the rest of the Uranian 'year' conditions are less extreme. It was thought that the axial rotation period was between 10 and 11 hours, a value now known to be considerably too short.

That, really, was about as much as we knew, and the results from Voyager 2 were eagerly awaited.

Voyager 2 was launched in 1977, a few days before its twin Voyager 1 (which, incidentally, went nowhere near Uranus or Neptune, and is now on its way out of the Solar System). In 1979 Voyager 2 by-passed Jupiter, and sent back spectacular pictures as well as masses of data. It then went on to a rendezvous with Saturn in 1981, again with excellent results, after which it swung out towards an encounter with Uranus. It was, in effect, playing a game of what I have irreverently termed interplanetary snooker, using the gravitational pull of one planet to send it out to the next — a procedure made possible by the fact that in

65

the late 1970s the four giant planets were arranged in a curve. This was lucky, because it will not happen again for well over a hundred years.

The power for the space-craft is supplied by what is in effect a tiny atomic generator (you cannot use solar power in those remote regions, because there is not enough sunlight), and the almost incredibly weak signals are picked up at various receiving stations, one at Goldstone in California, one in Spain and two in Australia. Everything is then sent through to Mission Control, at the Jet Propulsion Laboratory in Pasadena, California. This is where I was during the January encounter, together with scientists from all over the world.

The news media, of course, were well represented. I had been at Mission Control for all the previous planetary encounters — Mars, Jupiter and Saturn — but this time the feeling of excitement was perhaps greater than ever, because we knew so little about Uranus as a world. We also know that Voyager 2 was likely to be the last Uranus probe for a very long time.

The first major discovery came on 30 December 1985, almost a month before closest approach. The Voyager pictures showed a new satellite, now named Puck, closer in than any of the previously known satellites although still well outside the ring-system. Its diameter is about 105 miles. Calculations showed that on 'encounter day', 24 January, Voyager would pass within 320,000 miles of the new satellite, and the chance of taking a picture of it was really too good to be missed. Puck proved to be darkish and almost spherical, with three obvious craters — now given the rather curious names of Bogle, Lob and Butz.

A second new satellite (Portia) was found on 3 January, and others followed, until a total of ten had been discovered. All are small (Portia 67 miles in diameter, the others even less), and the two closest in, Cordelia and Ophelia, lie on opposite sides of the outermost ring of Uranus. Cordelia can be no more than about 16 miles across. Certainly there would have been no hope of discovering any of these miniature moons if we had been limited to using Earth-based telescopes.

As Voyager drew in, there was little to be seen on the pale, greenish disk of the planet itself. It was only a few days before closest encounter that the first clouds were seen. They were not conspicuous, and could not have been less like the vividly-coloured clouds of Jupiter or Saturn; but they existed, and they made it possible to measure the length of Uranus' axial rotation, which turned out to be $17\frac{1}{4}$ hours.

TITANIA, FROM 229,000 MILES Taken by Voyager 2, 24 January 1986. Note the prominent craters and the massive fault valleys.

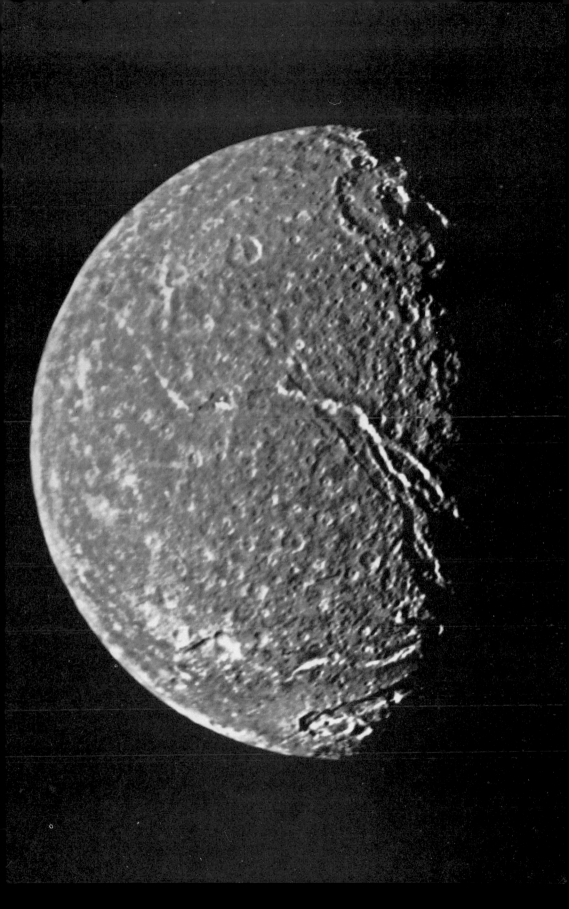

MIRANDA, SHOWING THE 'CHEVRON' AND THE TWO TYPES OF TERRAIN

The next revelation was that Uranus does indeed send out radio waves, and there is a magnetic field which is of considerable strength. Yet the magnetic axis is nowhere near the rotational axis; the two are separated by 60 degrees, for reasons which remain a total mystery. Moreover, the magnetic axis does not pass through the centre of the planet; it is appreciably offset. And to make matters even more complicated, the magnetic pole now facing the Sun (and the Earth) has north magnetic polarity, whereas it should logically be south. It is all very strange. Next came the discovery of the 'electroglow', unlike anything found on the other giant planets; it is a glow in Uranus' atmosphere which extends upward for at least 30,000 miles. Its mechanism is still not understood.

Though it is unwise to come to any definite conclusions, the Voyager results indicated that our ideas about the internal structure of Uranus needed revision. We had expected a rocky core, surrounded by a liquid ocean of melted ices (ammonia, methane and water), above which would come the atmosphere, with its ammonia and methane clouds. There is certainly a rocky core, but it now seems more probable that surrounding it is a layer in which the gases and ices are mixed in a dense layer, with the clouds above. The temperature at the surface of Uranus is so low that methane is able to condense above the other clouds.

The rings provided surprises of their own. Several new ones were found; the outermost ring is variable in width, and seems to contain pieces of darkish material which are several feet in diameter. Also, the whole of the Uranian ring system is pervaded by blackish 'dust' of unknown composition.

On 'encounter day', 24 January, all the previously known satellites were surveyed in rapid succession, and revelation followed revelation. The satellites are not identical. The two largest, Titania and Oberon — each about 1,000 miles across, less than half the diameter of our Moon — are icy and cratered; some of Oberon's craters have dark internal deposits, while Titania shows towering ice-cliffs. Umbriel and Ariel are rather smaller, but Umbriel is the fainter of the two, because it has a darker surface. Its craters are rather subdued, but there is one strange feature (now named Wunda) which is decidedly mysterious; it is relatively bright, and seems to be around 90 miles across, but it was right on the limb of the satellite as seen from Voyager, and we cannot be sure that it is a normal crater. Ariel is dominated by craters and by broad, branching, smooth-floored valleys which give every impression of having been cut by liquid. Since Ariel has an icy surface, it is tempting to conclude that the valleys must have been cut by water in the remote past — but Ariel is only 720 miles in diameter, with a very low escape velocity, so that it would not have been expected to have a dense atmosphere at any stage during its career.

Miranda, only 300 miles across, is the most bizarre of all. It has an incredibly varied surface; craters, scarps, grooves, ice-cliffs, chaotic terrain and two wide features which were originally referred to as 'the Chevron' and 'the Race-track', but are now more soberly known as

Inverness Corona and Arden Corona. Every type of feature seems to be present. As I commented at the time: 'You name it — Miranda has it!'

Certainly we had plenty to think about as Voyager 2 moved away from Uranus, and began the journey out to its final target, Neptune, which it is due to pass in August 1989. Old though it is, Voyager is still working even better than it did when encouterning Jupiter and Saturn. An on-board computer fault which developed as the space-craft neared Uranus was actually put right by a command sent from the Deep Space Network in Pasadena. Even if it does nothing more, Voyager 2 will still go down in history as one of the most triumphant probes of the first decades of the Space Age. But there is no reason to doubt that it will be equally successful at Neptune, and send back high-quality pictures as well as miscellaneous data. After that Voyager will move away into interstellar space, and eventually we will be able to track it no more.

So let us salute Voyager 2 and its makers. Finally, note that it is an excellent timekeeper. After a journey of so many millions of miles, and after having been in space for over eight years, it reached its closest point to Uranus precisely one minute nine seconds early. British Rail, please copy!

13 The Black Comet

Many people will remember seeing Halley's Comet during its period of visibility in 1985 and 1986, but by no means everybody was impressed by it. As we have noted, it was nothing like as brilliant as it had been in 1910, because it did not come so close to us, and when at perihelion it was on the far side of the Sun, so that we could not see it at all. Yet from a purely scientific point of view this was the most important return of all, because of the five space-probes: Suisei and Sakigake from Japan, Vega–1 and Vega–2 from Russia, and Giotto from Europe.

Remember that up to that time we had not really known what a cometary nucleus was like. The favoured theory, due to Fred Whipple, was that the only really massive part was the nucleus, graphically described as a 'dirty snowball', made up of ice — chiefly ordinary water ice — together with solid particles. When well away from the Sun the comet was a tiny, frozen mass, but as it came closer-in the ices would start to evaporate, surrounding the nucleus with a head or 'coma' and effectively hiding it. There could also be a tail or tails,

THE AUTHOR WITH FRED WHIPPLE AT DARMSTADT We were broadcasting during the Giotto encounter with Halley's Comet.

made up of gases driven away from the coma by the pressure of the solar wind, or dust, due to the pressure of sunlight. Yet there were still supporters of a rival theory, championed by the Cambridge astronomer, R.A. Lyttleton, who maintained that there was no icy nucleus, and that a comet was simply a flying gravel-bank made up of small particles which jostled together when reaching their closest point to the Sun.

Giotto, it was hoped, would clear the matter up. I had seen the launch, from Kourou in French Guyana; on 6 March 1986 I arrived in Darmstadt in Germany, headquarters of the European Space Agency, to report on the Giotto encounter with the comet a week later. I was particularly interested, in as much as I was also a member of the International Halley Watch, in my rôle as an astronomer rather than a television reporter, and I had been photographing and studying the comet as often as I could.

71

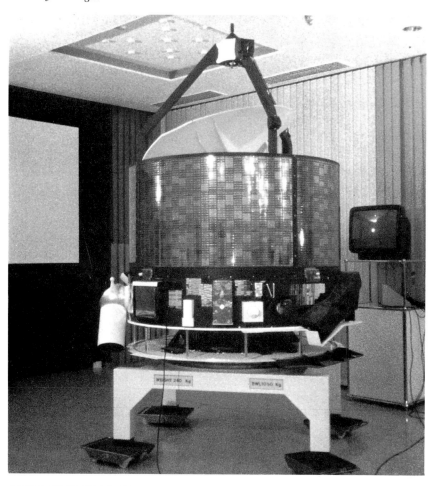

A FULL-SCALE MODEL OF GIOTTO, ON SHOW AT DARMSTADT DURING THE ACTUAL MISSION

Darmstadt was a hive of activity. There were arguments, too, because Giotto carried a great many different experiments, and the various teams had different requirements. Some of the experimenters wanted Giotto to go as close to the comet's nucleus as possible, while others preferred to stay farther out. In particular, the camera team, led by Dr H. Keller of West Germany, wanted to go no closer-in than 600 miles. Eventually a compromise was reached: just over 300 miles. The actual minimum range was 375 miles, and it transpired that Giotto arrived only four seconds late, after a journey which had started in the previous July. As a timekeeper, it was just as accurate as Voyager 2 had been in its journey to Uranus.

It was fairly obvious that the main danger to Giotto would come from the impacts of dust-particles. Halley's Comet orbits the Sun in a wrong-way or retrograde direction, so that Giotto had to meet it head-on at a relative speed of 43 miles per second — and at this velocity, even a tiny dust-particle has tremendous destructive power. Bumper shields, designed by Fred Whipple, had been fitted, but nobody knew how effective they would be. Whipple was there in person; I spoke to him a few hours before the encounter, and he remained confident that Giotto would survive. He expected to see an icy nucleus, with spurting jets and streamers. There was little support for the views of Sir Fred Hoyle and his colleague Professor Chandra Wickramasinghe, who claimed that the comet would prove to have a very dark nucleus coated with organic materials.

The first essential was to find out just where the nucleus was; it was completely hidden by the coma, and could not be seen at all. This was where the Japanese and the Russian space-craft helped. They by-passed the comet well before Giotto made its sortie, and their results were of the utmost value. Roald Sagdeev, the chief Russian investi-gator, was at Darmstadt in person, and actually joined me in the *Sky at Night* programme, together with Fred Whipple and also Jan Oort, the great Dutch astronomer who had first proposed the idea of a 'cloud' of comets at around a light-year from the Sun.

When we went 'on the air' on the late evening of 13 March there was a tremendous atmosphere of tension. Nobody knew what to expect, or even whether Giotto would survive the battering it was bound to receive as it plunged into the comet. In fact the first dust impacts were felt at a distance of 175,000,000 miles from the nucleus, and not until the distance had been reduced to 5,000 miles was there an impact violent enough to cause any alarm. But fourteen seconds before closest approach, when Giotto was still 1,058 miles from the nucleus, there was a direct strike from a piece of material about the size of a grain of rice. The space-craft was 'jolted' out of alignment, and the signals became intermittent. For a moment we feared that Giotto had been destroyed. Luckily, it had survived, but no pictures were obtained of the outgoing journey, so that our sight of the nucleus was confined to one side only.

I had wanted to present the entire *Sky at Night* programme from Darmstadt. However, this was beyond our budget, and it was decreed that the *Horizon* programme should become involved, putting on a discussion from Greenwich Observatory (the old Royal Observatory, that is to say; not Herstmonceux). I did not believe that this would be a success, and to be candid I think that subsequent events proved me right, but at least I was able to involve all the leading 'comet men' from Darmstadt, and by the end of the programme we had learned a great deal.

We had expected the main results to come from the HMC or Halley Multi-colour Camera, and we were not disappointed. There were some major surprises. First, the nucleus was much warmer than had been predicted. In shape it was likened variously to a baked bean, an avocado pear or even a banana; it was about 9 miles long and 5 miles wide, with a mass of from 50,000 million to 100,000 million tons. (This may sound a great deal, but it would need at least 60,000 million comets of this mass to equal the mass of the Earth.) The low density indicated that the comet was made up of 'fluffy' material, with water ice as the main constituent, together with other substances such as carbon dioxide, formaldehyde, nitrogen and ammonia. There was considerable surface structure, including a bright region which seems to have been a hill almost a mile high; there were craters, and jets, and three main areas of activity. Yet all the jets seemed to issue from one small area on the sunlit side of the nucleus, and the remaining 85 per cent of the surface was inactive. The rotation period was given as 7.3 hours, and there was a 'precessional' period of 53 hours, so that the nucleus was behaving rather in the manner of a spinning gyroscope. It was also estimated that the comet would have lost about 300,000,000 tons of material by the time that it moved back into the remoter parts of the Solar System and would become inert once more.

But the real shock was the colour of the nucleus. It was not ice-bright, as most people had expected; it was black. Dr Keller compared it with 'black velvet', reflecting no more than 2 to 4 per cent of the sunlight falling upon it. There were two bright patches, each about $1\frac{3}{4}$ miles across, which had been recorded by the Russian Vegas and had given the misleading impression of a double nucleus, but the general aspect was black.

What is the explanation? There must be a layer of dark dust which protects the ice below from evaporating quickly and is an effective insulator. It seems, then, that Whipple's 'dirty snowball' contains more dirt than he had expected.

This brings us back to the theories of Fred Hoyle and Chandra Wickramasinghe, who were certainly correct in saying that the nucleus would be dark, and may well have been correct also in maintaining that organic materials were very much in evidence. (In fact, Wickramasinghe telephoned me at Darmstadt a few hours before the Giotto encounter, and said precisely this.) They have also claimed that comets are bringers of life; that living things on Earth were brought here by way of a comet; that comets produced our oceans, and that they can deposit viruses in our atmosphere, thereby giving rise to epidemics. Ideas of this sort have been received with profound scepticism, particularly by medical specialists, but it is undeniably fair to say that

HALLEY'S COMET This was taken with the UK Schmidt Telescope (Siding Spring) on 13 December 1985.

Hoyle and Wickramasinghe have won the first round.

Long before Halley's Comet comes back, in 2061, there will have been other cometary probes, some of which are already in an advanced stage of planning. Whether they too will show black nuclei remains to be seen, but there is no reason to suppose that Halley's Comet is anything but typical. Meanwhile, the story of Giotto itself is not over; it has been badly damaged, and the camera is permanently out of action, but there are hopes that it may eventually be salvaged, and it is still under our control as it makes way back towards Earth. It will never be forgotten; in a few hours on the night of 13 March 1986 it told us more about comets than we had learned throughout the whole history of science — and remember, it was essentially a British space-craft.

14   The Story of Saturn

Of the five planets known in ancient times, Mercury and Saturn are much the least imposing. Mercury is never conspicuous, simply because it keeps close to the Sun; Saturn may be visible throughout the hours of darkness, but shows up as nothing more than a moderately bright, slightly yellowish star. It is a slow mover, and the old star-gazers named it after the gloomy original ruler of Olympus. Astrologers regarded it as baleful, and we still use the adjective 'saturnine'. In pre-telescopic times there was no way of telling that this uninspiring object is in fact the most beautiful object in the entire sky.

The first man to observe Saturn through a telescope was of course Galileo. In 1610 he completed his primitive 'optick tube', and turned it skyward. He saw the craters of the Moon, the satellites of Jupiter, the phases of Venus and the countless stars of the Milky Way, but Saturn was a puzzle. There was something unusual about it, but the definition given by his telescope was not good enough for him to decide what it was. He came to the conclusion that Saturn must be a triple planet, with a large central body and smaller ones to either side. This was strange enough, but two years later Galileo was even more perplexed; the two minor bodies had disappeared. Galileo wrote:

'Have they vanished or suddenly fled? Has Saturn, perhaps, devoured his own children? . . . . I do not know what to say in a case so surprising, so unlooked-for and so novel. The shortness of the time, the unexpected nature of the event, the weakness of my understanding, and fear of being mistaken, have greatly confounded me.'

We now know the answer. Saturn's magnificent rings are extensive, measuring almost 170,000 miles from one side to the other, but they are also very thin, so that when they are turned edgewise-on to us they disappear in small telescopes; even with large instruments they are difficult to follow, and appear only as an excessively faint line of light.★ Galileo, predictably, lost them during the edgewise presentation of 1612. Later he saw them again, but he failed to interpret them, even though one of his drawings — made in 1616 — was surprisingly

---

★At the last edgewise presentation, that of 1980, I was using the 24-inch refractor at the Lowell Observatory in Arizona, one of the best telescopes of its kind in the world. I did not actually lose the rings, but even under good conditions I found them excessively difficult.

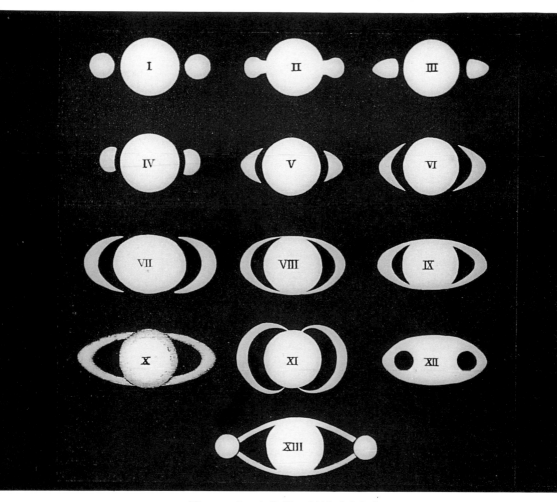

EARLY DRAWINGS OF SATURN They are by: 1 Galileo 1610. 2 Scheiner 1614. 3
Riccioli 1641. 4–7 Hevelius c. 1645. 8, 9 Riccioli 1648–50. 10 Fivinim 1646–8.
11 Fontana 1636. 12 Bianchini 1616, Gassendi 1638–9. Fontana 1644–5.

accurate, and showed the 'minor bodies' as handles rather than as
separate globes.

During the early part of the seventeenth century telescopes were
much improved, though they would have looked strange to modern
eyes; Christiaan Huygens, the best observer of the time, used a small-
aperture refractor with a tube a dozen feet long. It could give a
magnification of 50, and with it Huygens discovered Saturn's brightest
satellite, Titan. He also realized that Saturn was not triple, but was
'surrounded by a flat ring, which nowhere touches the body of the

SATURN, FROM VOYAGER 2

planet'. Strangely, some of his contemporaries were reluctant to accept this idea. The French mathematician de Roberval believed Saturn to be surrounded by 'a torrid zone giving off vapours'; Hodierna, of Sicily, claimed that Saturn was merely a globe with two dark patches on it; Honoré Fabri, a Jesuit professor of philosophy, introduced two dark satellites and two bright ones. Sir Christopher Wren who, remember, was a professional astronomer before turning to architecture, had his own theory; but before he could publish it he heard Huygens's account, and at once accepted it. It was not until 1665 that the ring theory was finally proved.

It was also at this time that G.D. Cassini, in Italy, discovered the main gap in the ring-system which is still called the Cassini Division. A minor division in the outer ring was found in the nineteenth century by Johann Encke, and in 1850 came the detection of the inner 'dusky ring' closer to the planet. The ring system was, apparently, fairly uncomplicated; two bright rings, one fainter ring, and two principal divisions. But what about the planet itself? Could it be a minor sun, providing warmth for its satellite system — of which eight members were known by 1850?

Many astronomers believed so. In 1882 Richard Proctor, in his famous book about Saturn, wrote as follows:

'Over a region of hundreds of thousands of square miles in extent, the glowing surface of the planet must be torn by subplanetary forces. Vast masses of intensely hot vapour must be poured forth from beneath, and rising to enormous heights, must either sweep away the enwrapping mantle of cloud which had concealed the disturbed surface, or must itself form into a mass of cloud. . . .'

It was a fascinating picture, and it seemed plausible enough, but we now know that it is completely wrong. The outer clouds of Saturn are not hot; they are bitterly cold. True, the core is at a high temperature, but Saturn cannot send out any perceptible heat to its satellites. We now believe that there is a rocky core, overlaid by deep layers of liquid hydrogen which are in turn overlaid by the gaseous atmosphere. The huge globe, over 70,000 miles in diameter, is quite unlike that of the Earth, and its overall density is actually less than that of water. It has been said that if Saturn could be dropped into a vast ocean, it would float.

Three space-craft have now by-passed Saturn, and all our long-cherished ideas have had to be drastically modified. Unfortunately, the *Challenger* Shuttle disaster has held up the American programme of research, and the Russians have not yet tried to send any probes into the outer part of the Solar System. The next mission to Saturn and its system will be named in honour of Cassini; it should have been launched early in the 1990s, but it cannot now hope to make its pass until well into the twenty-first century. Until then, we must be content with analyzing the data which we already have.

Though both Jupiter and Uranus are now known to have rings, the system of Saturn is in a class of its own. The rings are beyond the range of binoculars, but even a small telescope will show them, together with several of the satellites. During the mid and late 1980s the rings are wide open as seen from Earth, so that Saturn appears at its very best. There are few people who, on seeing the rings for the first time, do not give a gasp of wonder and admiration.

15   Globes of Stars

Edmond Halley is always remembered mainly as the man who first predicted the return of the famous comet which has just paid us its latest visit. Yet this discovery was only one of many which Halley made during the course of his long career. He was what we would nowadays call an all-rounder, and his interests ranged from astronomy to archaeology, meteorology, undersea diving, ancient history and even military fortifications. Astronomy, of course, was his main love (he was Astronomer Royal for more than twenty years), and he was an expert observer.

In 1714 he looked at a region in the rather dim constellation Hercules, which is high above the British horizon during the evenings in summer and through to autumn. There, between the stars Zeta and Eta Herculis, he saw what he called 'a little patch' which 'shows itself to the naked eye when the sky is serene and the Moon absent'. He did not then realize that he was looking at a vast system of stars — the first recorded object of the type we now call a globular cluster.

Globular clusters are quite different from loose or open clusters such as the Pleiades. Whereas the open clusters are quite formless, and often rather sparse, globulars are symmetrical and rich. A large cluster such as that in Hercules may contain more than a million stars, many of them much larger and more luminous than our Sun; near the centre of the system they are so closely packed that even in a powerful telescope they seem to merge into one mass. In a way this is misleading, because even in the densest part of a globular cluster the stars are still widely separated, and collisions must be rare. However, the whole region is much more crowded than our part of the Galaxy, and the night sky to an observer inside a globular cluster would indeed be glorious.

The Hercules cluster is on the fringe of naked-eye visibility, as Halley had pointed out. It lies directly on a line joining Zeta Herculis to Eta, rather closer to Eta. Binoculars show it very clearly, and even a small telescope will resolve the outer parts into stars. It is one of the finest of all globulars; only Omega Centauri and 47 Tucanae are brighter, and both these are too far south to rise over Britain. The Hercules cluster is known officially as Messier 13, or M.13, because it was the thirteenth object in a catalogue of clusters and nebulae drawn up by the French astronomer Charles Messier in 1781. (Omega

Centauri and 47 Tucanae have no Messier numbers, because from his home in Paris Messier could never see them.)

If M.13 contains a million stars, which is probably a good estimate, then it must be a long way away. The distances of globular clusters are not easy to measure, but there is one method which we believe gives reliable results. It involves using stars which are not constant in light, but which fluctuate regularly over short periods. These Cepheids, named after the prototype star Delta Cephei in the far north of the sky, have periods of from a few days to a few weeks, and are highly luminous; it has been found that their periods are directly linked with their real output, and there is a definite relationship stating that 'the longer the period, the more powerful the star'. There are also the so-called RR Lyrae variables, which have periods of less than a day, and all of which seem to be about 100 times as luminous as the Sun.

The discovery of the Cepheid period-luminosity law was made just before the First World War by Henrietta Leavitt, working in America. She was studying photographs taken which showed the Small Cloud of Magellan, which again is in the far south of the sky, but is one of the nearest of the systems we now know to be external galaxies (it is less than 200,000 light-years away). Miss Leavitt found many Cepheids in the Small Cloud, and realized that to all intents and purposes they could be regarded as being at the same distance from us, just as it is usually good enough to say that Victoria and Charing Cross are the same distance from New York. The Cepheids with the longer periods looked the brighter, and so they were genuinely the more luminous. There was no reason to doubt that the Cepheids in the Small Cloud were different from Cepheids anywhere else, and the period-luminosity law would also apply to Cepheids in our own part of the universe.

A few years later Harlow Shapley, at Mount Wilson, set out to measure the size of our star-system or Galaxy. He knew that the globular clusters were very remote, and that they probably lay around the edge of the main Galaxy; he also knew that they are more numerous in the Southern Hemisphere, particularly in the constellation of Sagittarius (the Archer), so that we were having a lop-sided view. This, he reasoned, was because the Sun lay well away from the galactic centre. He detected short-period variables in the globular clusters, and by measuring their periods he could find their real luminosities, after which he could work out the distances of the globular clusters inside which they lay. According to modern values, the Hercules globular is between 21,000 and 24,000 light-years away from us. The overall diameter of the Galaxy is of the order of 100,000 light-years, and the distance between the Sun and the galactic centre is slightly less than 30,000 light-years, though obviously we cannot be

THE SOUTHERN GLOBULAR CLUSTER 47 TUCANAE Photographed by Alan Gilmour at the Mount John Observatory, New Zealand.

precise; there are many complicating factors, notably the absorption of light by the interstellar material which is spread thinly all through space.

Globular clusters are very ancient by cosmical standards. Stars of around the same mass as the Sun begin by condensing out of the gas-and-dust clouds which we call nebulae, and shine by nuclear reactions going on deep inside them. Their main 'fuel' is hydrogen; and when the hydrogen starts to run low the star has to readjust itself. The outer layers expand and cool, turning the star into a red giant. The leading stars in globular clusters have already reached the red giant stage, so that they are highly evolved.

Northern Hemisphere observers regret that the two finest globular clusters are inaccessible to them. Omega Centauri is a very conspicuous naked-eye object, and is easy to find, not too far from the brilliant Pointers to the Southern Cross. A modest telescope will resolve much of it. Hardly inferior to it is 47 Tucanae, in the least conspicuous of the 'Southern Birds' and fairly close to the south celestial pole. It too is easy to see with the naked eye, but by sheer coincidence it lies almost in front of the Small Cloud of Magellan, and it is tempting to think that the two are associated. Of course, they are not; 47 Tucanae is a member of our Galaxy, while the Cloud is external, and at least ten times farther away. If you look at them with binoculars you will see that the apparent surface brightness of 47 Tucanae is much greater than that of the background Cloud.

Telescopically, I always think that 47 Tucanae is the more impressive of the two. Except with a very low magnification, Omega Centauri will more than fill the telescopic field, and the full glory will be lost, whereas the smaller 47 Tucanae can show up as a complete system.

Of course, there are other globulars within the range of small telescopes or even binoculars, and some of them, such as M.92 in Hercules and M.5 in the Serpent, are close to naked-eye visibility. But in the northern hemisphere of the sky, M.13 is the best example, and it is unusual only inasmuch as it contains fewer short-period variables than might have been expected. Its overall diameter is about 160 light-years.

At the centre of M.13 the stars must be only light-weeks or even light-days apart, as against several light-years for our part of the Galaxy. To an inhabitant of a planet moving round a star deep in M.13 there would be no proper darkness, because many stars would be brilliant enough to cast shadows, and many of these stars would be red. Whether any life exists there we do not know, but during the 1970s the giant radio telescope at Arecibo, in Puerto Rico, was used to beam a message out to the cluster. If the message is safely received, and a reply sent at once, we may hope for some response around the year AD 40,000!

---

16 The Death of a White Dwarf

More than ten million years ago there was a violent outburst in the galaxy which we call Centaurus A. It was a supernova — a colossal stellar explosion, resulting in the complete destruction of a star — but we knew nothing about it until 1986, when the supernova was discovered by an Australian amateur astronomer, the Rev. Robert Evans. Since then it has been closely studied by professional astro-

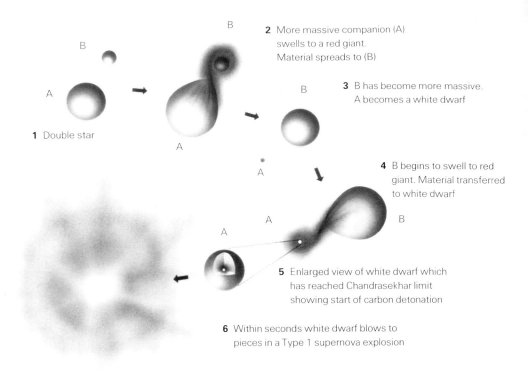

1 Double star

2 More massive companion (A) swells to a red giant. Material spreads to (B)

3 B has become more massive. A becomes a white dwarf

4 B begins to swell to red giant. Material transferred to white dwarf

5 Enlarged view of white dwarf which has reached Chandrasekhar limit showing start of carbon detonation

6 Within seconds white dwarf blows to pieces in a Type 1 supernova explosion

SUPERNOVA DEVELOPMENT

nomers who are able to see it; unfortunately, it is too far south in the sky to rise over Britain or the northern United States.

Centaurus A is a curious system. It was once believed to be made up of two galaxies which were 'passing through' each other, much in the manner of two orderly crowds walking in opposite directions, but it is now known to be a single galaxy, crossed by a dark dust-lane which gives it a most unusual appearance. It is not a member of our Local Group of galaxies (a collection which includes the Milky Way system, the Magellanic Clouds, the Andromeda and Triangulum spirals and more than two dozen smaller systems), but its distance, which has been given as 13,000,000 light-years, makes it one of our nearer neighbours. This is why the outburst of a supernova in it is so important.

A supernova occurs when a formerly very faint star flares up to

THE CRAB NEBULA This is the most famous of all supernova remnants. The photo was taken with the Palomar 200-in reflector.

many times its normal brilliancy, remaining prominent for a few days, weeks or even months before fading away. Supernovae are quite different from ordinary 'new stars' or novae, and are much more powerful, reaching peak luminosities hundreds of millions of times that of the Sun. However, they are not common. During the last thousand years only four have been definitely observed in our Galaxy: the stars of 1006 (in Lupus, the Wolf), 1054 (in Taurus, the Bull), 1572 (in Cassiopeia) and 1604 (in Ophiuchus, the Serpent-bearer). All these became brilliant enough to be seen with the naked eye in broad daylight, and the 1006 supernova may have rivalled the quarter-moon, though unfortunately it was not well documented.

The supernova of 1054 was studied by the Chinese, and has left the remnant which we call the Crab Nebula, a patch of expanding gas in the midst of which is a tiny, amazingly dense, rapidly-spinning object known as a pulsar; it is made up of neutrons, and is the Crab's 'power-house'. In 1572 another supernova was followed by the great Danish astronomer Tycho Brahe, and in 1604 we had the observations of Johannes Kepler, so that this particular supernova is always known as Kepler's Star (even though he was not the actual discoverer). Other supernova remnants are known, and one of these, Cassiopeia A, is a radio source associated with a ring of gaseous material which is expanding outward from the old explosion-centre. Probably the outburst would have been around the year 1667, but it lay in an area which is obscured by interstellar dust, so that it was not observed with any certainty. It was unquestionably what is termed a Type I supernova, and here we must differentiate between the two types, because they are very different in nature.

A Type II supernova includes the virtual destruction of a massive star, which 'runs out of fuel' and explodes, leaving an expanding gas-cloud together with either a pulsar or a black hole. The Crab supernova was of this type. I will have more to say about Type II supernovae later, so for the moment let us concentrate upon those of Type I.

Here we begin with a binary system, made up of two stars which are moving together round their common centre of gravity. One star (A) is much more massive than its companion (B). The more massive a star, the more quickly it evolves, so that A runs through its life-story more quickly than B. Eventually it swells out to become a red giant, and material is pulled across from it to B, so that B grows in mass while A declines. When this has continued for a sufficient length of time B has become the more massive of the two, while all that is left of A is a very small, very dense core, made up chiefly of carbon and known as a white dwarf.

Now the situation is reversed. B evolves in its turn, swells out, and starts losing its material back to the shrunken A, which is still pulling strongly. Steadily the white dwarf builds up a gaseous layer, composed largely of hydrogen which has been stolen from B. But, as was shown years ago by the Indian astronomer S. Chandrasekhar, there is a limit beyond which a white dwarf cannot go. If its mass becomes more than 1.4 times that of the Sun, disaster will overtake it, and with a supernova this is exactly what happens. When the mass of A goes over the limit, the carbon detonates, and the star blows itself to pieces in a matter of seconds. The power generated is absolutely staggering, and the outburst takes a long time to die away.

No supernova in our Galaxy has been observed since 1604, and this was before the invention of the telescope. However, supernovae are so luminous that we can locate them in other galaxies, and many have now been seen, most of them many thousands of millions of light-years away. Amateur astronomer have now joined in the hunt for them, and make systematic searches, often with highly sophisticated equipment. For instance, R.W. Arbour, in England, has a fully-computerized 16-inch reflecting telescope which can swing automatically from one galaxy to another, taking a five-minute photographic exposure of each; when the photographs are developed, any unusual object such as a supernova betrays itself at once.★ The method used by the Rev. Richard Evans, in Australia, is different. He searches visually, using

★Arbour was also the second astronomer in Britain to photograph Halley's Comet in the summer of 1985, when the comet emerged from the rays of the Sun. He was beaten by about fifteen minutes by another amateur, Alan Young of Sussex. I was clouded out from Selsey, but John Mason and I managed to photograph the comet on the following night, using my 15-inch reflector, and a few hours later it was also photographed from the Royal Greenwich Observatory in Herstmonceux.

THE GALAXY CENTAURUS A, WITH THE SUPERNOVA ARROWED

his $12\frac{1}{2}$-inch reflector for much of the work, and he has memorized the aspects of his target galaxies so perfectly that he can recognize anything unusual without delay. It was so with the supernova in Centaurus A.

The fact that Evans reported it so early meant that it could be followed professionally from an early stage in its outburst, and a light-curve drawn up. A Type II supernova blazes forth quickly and then declines fairly smoothly; a Type I increases just as quickly, but then drops in brightness before the light-curve shows a long 'tail'. The behaviour of the Centaurus supernova (known as SN 1986G, because it is the seventh supernova to have been seen this year) proves that it is of Type I. Moreover, additional evidence is always to hand with a star of this kind. When the carbon white dwarf explodes, it creates other elements including neon, oxygen and silicon, ending up as nickel. Nickel decays to cobalt — it takes six days for half of it to be converted — and then to iron, in a 'half-life' of 77 days. Sure enough, the spectrum of 1986G soon showed the tell-tale signs of iron.

It is fortunate that the supernova was on the outer edge of the dust-lane in Centaurus A; had it been deeply embedded, we would have been unable to see it. Even so, its light has come to us via a layer of dust, and it appears redder than it would otherwise do, so that it can tell us a great deal about the nature of the dust in Centaurus A itself.

This is only part of the information which can be drawn from the spectrum of the supernova. There are traces of clouds due to the element calcium, some of which are genuinely in Centaurus A and which show that there are great velocity differences of the clouds over the galaxy, finally disposing of the old idea that Centaurus A is a combination of two systems in collison. Moreover, we can trace clouds of calcium lying between Centaurus A and our own Galaxy, the origin of which are still decidedly uncertain. Similar effects were seen in 1983 in a supernova which flared up in the spiral galaxy Messier 83, in the same part of the sky, and no doubt the same intergalactic calcium cloud is responsible.

Because Nature has not been obliging enough to provide us with a bright supernova in our Galaxy since 1604, we have to make do with those in outer systems, so that the work of amateurs is particularly valuable. Of course, professional astronomers make systematic searches — first Fritz Zwicky, and more latterly Charles Kowal at Palomar — but they cannot monitor all the available galaxies all the time. Evans has now made over a dozen discoveries, and will no doubt make many more with his original reflector and his newly-acquired 16-inch. When he addressed the 1986 Congress of the International Astronomical Union he was given a standing ovation, and nobody could have deserved it more.

17 Missions to Mercury

In August 1986 I travelled to Tucson, in Arizona, to present a paper at an international conference dealing with the planet Mercury. The meeting was interesting, and was well attended by astronomers from many countries; its main object was to 'pull together' all that we have discovered about the innermost member of the Sun's family.

Mercury is never very conspicuous. It orbits the Sun at a mean distance of only 36,000,000 miles, as against 93,000,000 for the Earth, so that to us it always seems to stay close to the Sun in the sky; with the naked eye it is visible only when best placed, either low down in the west after sunset or low down in the east before sunrise. It is actually quite bright, and can outshine all the stars apart from Sirius, but it cannot be observed against a dark background, and there must be many people who have never seen it at all. Yet it was well known in ancient times, and was named in honour of Hermes or Mercury, the elusive, fleet-footed Messenger of the Gods.

To make matters even more awkward, Mercury is a small world, just over 3,000 miles in diameter and therefore not a great deal larger than the Moon. It never comes much within fifty million miles of us, and when at its closest it is virtually between the Sun and the Earth, so that its dark or night side is turned towards us. There are occasions when Mercury is exactly lined up with the Sun, so that it is seen as a black spot passing across the solar disk from one side to the other, taking a few hours to do so. These transits are not common. The last was on 13 November 1986, though it was not visible from England because the Sun was below the horizon; the next transit will not occur until 6 November 1993.

A small telescope will show the phases of Mercury, from crescent to nearly full, but little else. Surface markings are very hard to make out. The first attempt at a map was made by the Italian astronomer Giovanni Schiaparelli more than a century ago; he was followed by others, notably by Eugenios Antoniadi, who used the powerful 33-inch refractor at the Paris Observatory. The two maps showed dark and brighter areas which were vaguely alike — but did they really represent genuine features?

Both Schiaparelli and Antoniadi made their observations when Mercury was high in the sky. As we have seen, Mercury and the Sun always stay fairly close together (the angular separation between the

A RAY-CRATER ON MERCURY Taken from Mariner 10. This crater has been named Kuiper in honour of the late G. P. Kuiper, a pioneer of planetary exploration by unmanned probe.

two can never be as much as 30 degrees), so that when Mercury is high the Sun is high too; but this cannot be helped. Both observers were sure Mercury spins on its axis in 88 Earth-days. This is also the time taken for the planet to complete one journey round the Sun, so that one area of the surface would have been in permanent daylight and another part in perpetual night. It was only in the 1960s that the dark side of Mercury was found to be much warmer than it would have been if it had never received any sunlight. The true rotation period is 58.6 Earth-days, or exactly two-thirds of a Mercurian 'year', which leads to a very peculiar calendar indeed.

In November 1973 the space-craft Mariner 10 was launched from Cape Canaveral in Florida. In the following February it by-passed the planet Venus, sending back good pictures of the upper clouds of that beautiful if decidedly lethal world; it then went on to a rendezvous with Mercury. It by-passed Mercury on 29 March 1974, and then entered an orbit round the Sun, again approaching Mercury on 21 September 1974 and on 16 March 1975. By then its instruments had started to fail, and all contact was finally lost on 24 March. No doubt Mariner 10 is still circling the Sun and still making regular close approaches to Mercury, but we have no idea of exactly where it is now.

Yet it was extremely successful, and has given us almost all our information about what Mercury is really like. The surface is mountainous and cratered, and at first glance looks very much like that of the Moon. The atmosphere is so thin that to all intents and purposes we can say that Mercury is an airless world. There is a dense, iron-rich core, and Mercury, unlike Venus or Mars, has a detectable magnetic field, though it is much weaker than that of the Earth. The maximum surface temperature is well over 700 degrees Fahrenheit, and any form of life there is absolutely out of the question.

Unfortunately, the same regions of Mercury were in sunlight during all three active Mariner passes, so that there is still more than half the planet waiting to be explored. Perhaps the most important feature discovered is a huge, ringed basin which has been named Caloris Planitia, or the Caloris Basin. Only half of it has been recorded — the other half was in darkness during the Mariner 10 mission — but we know that it is large, with a diameter of well over 800 miles, and that it is bounded by rings of smooth mountain blocks together with ridges. It has been so named because it marks one of the two 'hot poles'.

Mercury, unlike the Earth, has a somewhat eccentric orbit. At its closest to the Sun (perihelion) it is only 29,000,000 miles out; at its farthest (aphelion) about 43,000,000 miles. Near perihelion, an observer standing inside the Caloris Basin would see the Sun almost at the zenith or overhead point, and the temperature would be at its highest.

The sequence of events would be strange. At sunrise, the Sun would have its minimum apparent diameter. Slowly it would climb towards the zenith, growing in size; but when near the overhead point it would seem to stop, backtrack for eight Earth-days and then resume its normal motion, shrinking as it dropped towards the horizon and finally setting after an interval of 88 Earth-days. The time between one sunrise and the next would be 176 Earth-days. To an observer on Mercury's surface 90 degrees away from Caloris, things would be different. The Sun would seem largest when rising or setting, and it would appear briefly over the horizon at dawn before temporarily setting again; the same sort of effect would be seen at sunset. The reason for this curious behaviour is that while Mercury rotates at a steady speed, it moves quickest in its orbit when closest to the Sun, in accordance with the invariable traffic laws of the Solar System.

There is still a great deal about Mercury that we do not know, and it is tantalizing to have to rely almost solely upon a single space-probe which has been dead now for well over a dozen years. We still have to map more than half the surface; we do not know a great deal about the magnetic field, and we are uncertain about the nature of the extremely tenuous atmosphere. At the Tucson conference in 1987 all these problems were discussed, and it was agreed that our only hope of learning more was to send up a new space-probe.

The trouble, of course, is that the NASA planetary programme is still in a state of total disarray. Voyager 2 will by-pass Neptune in 1989, and luckily it cannot be stopped, but there are no schedules for future launchings to the outer planets, and even with the inner planets Mercury comes low in the list of priorities. One paper, by Dr J.R. French, outlined a novel method of travel: using a 'solar sail', a thin reflecting surface which would be propelled by solar radiation, and would therefore be very economical of power. The orbit of a solar-sail probe could be controlled by changing the orientation, much as a yacht can be made to sail towards the direction from which the wind is coming. With an interplanetary craft of this nature the acceleration would be very gradual, but the idea is by no means far-fetched. (It was even suggested years ago for the ill-fated American probe to Halley's Comet, and was rejected simply because there was not enough time to develop it.)

Of course, we have no positive knowledge about what the Russians may be planning, but it looks as though we must wait patiently for any new mission to Mercury, and the wait could be a long one. This is a pity — because hostile and lifeless though it may be, Mercury is a fascinating little world. Manned flight there is unlikely in the foreseeable future, but at least we may hope to receive pictures and information from an automatic probe which may well land in the scorched, desolate Caloris Basin.

18 The Furthest Depths of the Universe

'How big is the universe?' This is a question which has been asked time and time again, but which has never been answered. We cannot claim that we know the answer even yet, but at least we can probe out to incredible distances, and astronomers have just identified the most remote object known to science. It is a quasar, known by its catalogue number of Q0051–279, and it seems to be so remote that its light takes at least 13,000 million years to reach us.

The story of quasars really goes back to 1963, with some pioneer work by radio astronomers. The science of radio-astronomy deals with long-wavelength radiations (natural, not artificial, I hasten to add!), and has proved to be an essential branch of modern science. But in the early days, astronomers were puzzled to find that the most powerful radio sources in the sky were not brilliant stars such as Sirius and Vega, but were associated with much dimmer objects. The position of one source (known by its catalogue number of 3C–273) was tracked down, and identified with what had always been thought to be a faint, slightly bluish star. When the light from the 'star' was examined by means of a spectroscope, it was found that the object was not only remarkably luminous, but also remarkably remote; it seemed to be about a thousand million light-years away, so that we were seeing it as it used to be a thousand million years ago. This was the first quasar, a convenient abbreviation of the original but decidedly clumsy 'quasi-stellar radio source', or QSO. (Today, when we know that by no means all quasars are radio emitters, the term QSO is starting to come back into favour.)

The distance of 3C–273 was estimated by the familiar Doppler effect. When a source of light is moving away from us the light appears slightly 'too red', because fewer light-waves per second are reaching us than would be the case if the source were standing still, and the wavelength is effectively lengthened so far as we are concerned. The actual colour-change is slight, but the effect shows up in the spectrum. With a star, the spectrum produced when the light is split up by means of a prism, or some equivalent device, consists of a rainbow band, from red at the long-wave end down to violet at the short-wave end, crossed by dark lines; each line is due to some particular substance. If the star is receding, all the lines will be moved over to the long-wave end of the band, and the amount of this 'red shift' is a key to the velocity of recession.* The same is true of a galaxy, which is made up of thousands of millions of stars, and it could also apply to a quasar. With 3C–273, it was found that the red shift was very large indeed, leading to a vast distance, and, hence, a very high luminosity.

Astronomers were faced with the problem of explaining how it were possible to pack so much energy into an area which seemed to be no larger than our humble Solar System. By now we have found the answer — at least, so we hope. It is thought that quasars are the centres of very active galaxies, and that their energy is either gravitational or else due to a super-massive black hole. But what about their speeds?

For over half a century we have known that all the galaxies, except

---

* Will you forgive a mathematical formula? The red shifts of galaxies are usually denoted by z. Let L = the observed wavelength of the line in the spectrum of the galaxy, and let Lo = the wavelength which it would have if there were no red shift. Then $z = \frac{(L - Lo)}{Lo}$. For the quasar Q0051–279, z = 4.43.

those in our Local Group, are moving away from us. This does not mean that we are particularly unpopular; the entire universe is expanding. Moreover, the velocity of recession increases with distance. The same law applies to quasars, which are much more powerful than ordinary galaxies and can therefore be seen over a much greater range. (As we have noted, there are some astronomers who challenge this interpretation, but for the moment let us adopt the majority view.) 3C–273 was the first quasar to be identified, but others soon followed, apparently more remote and more luminous still. In 1972 one quasar was estimated to have a distance of 10,000 million light-years. This record was broken again and again, and by now, with Q0051–279, we have reached out to a distance of some 13,000 million light-years and a recessional velocity of 93 per cent that of light — that is to say, around 173,000 miles per second.

Obviously, we are approaching a critical distance at which an object (be it a galaxy or a quasar) will be speeding away at the full 186,000 miles per second velocity of light, in which case we will be unable to see it; we will have reached the boundary of the observable universe, though not necessarily of the universe itself. Unless Arp, Hoyle and their supporters are right after all, we are not so very far from the limit.

When we look at a remote object, we are also looking backward in time, and we see Q0051–279 as it used to be in the early history of the universe. The critical distance, beyond which we cannot see, may be anywhere between 15,000 million and 20,000 million light-years, with a slight preference for the lower figure.

This brings us on to another fundamental problem, that of the origin of the universe itself. The general view today is that all matter originated in a 'big bang', and that the overall expansion has continued ever since; but how the material was created in the first place remains a total mystery. As scientists, we may be strong on detail, but we are woefully weak on fundamentals. Neither can we say just 'where' the big bang occurred. If space, time and matter were created at the same moment, then the big bang happened 'everywhere'.

Also, will the present expansion continue indefinitely, until all the groups of galaxies and quasars lose touch with each other? Or will the present phase of expansion be followed by a period of contraction, until all the systems come together again in what might be called a 'big crunch', to be followed by a new phase of expansion? This depends upon the overall density of material in the universe, which may or may not be above the critical value. This is where the quasars may help us. If we can reach out to near the theoretical limit, we may be able to decide whether or not the expansion is slowing down significantly. In other words, we might find out whether or not there is enough material in the universe to pull the galaxies back.

Certainly there is not enough observed material to make this possible, but there is a great deal of hidden mass for which we cannot account, and the whole problem remains completely open. The missing mass might be locked up in black holes; it could be due to millions upon millions of dwarf stars too faint to be seen; it might even be caused by some type of material which we do not know how to detect — the possibilities are endless.

In any case, the size of the universe is something which we cannot appreciate. If it is finite, then we are entitled to ask what lies beyond, while if it is infinite we have to try to picture something which has no end. The language of mathematics indicates that the universe may be 'finite but unbounded', though this is almost impossible to put into words. (I once asked Albert Einstein whether he could do so. He admitted that he couldn't.)

Meanwhile, efforts continue to extend our range farther and farther. How long Q0051–279 will hold the distance record remains to be seen; I doubt whether it will be for long. Whether we will be able to reach the boundary of the observable universe remains to be seen, but there is at least a chance. The strange, super-luminous quasars give us cause for hope.

19 The Cosmic Lawn-sprinkler

Many strange objects have been found in the sky in recent times. One of the most remarkable of all is known as SS433, in the constellation of Aquila (the Eagle). It has caused a tremendous amount of interest among astronomers, and merits its name of the Cosmic Lawn-sprinkler.

It is not bright; it looks like nothing more than a dim star, far below naked-eye visibility, and until a few years ago it aroused no interest whatsoever. Then, quite unexpectedly, some significant facts began to emerge.

The story began with the examination of a gas-cloud which was catalogued as W.50, and which appeared to be the remnant of a supernova. Supernovae, as we have noted, are of two types. Type I involves the total destruction of a white dwarf star. In a Type II outburst, a very massive star runs out of fuel; there is an 'implosion' — the opposite of an explosion — and the star blows most of its material away into space, leaving a very small, amazingly dense core made up of neutrons. The density of a neutron star may be as much as a thousand million million times that of water.

The Crab Nebula, in Taurus (the Bull), is the remnant of a supernova seen in the year 1054. In its midst is a neutron star, which is spinning round rapidly and sending out pulsed radio waves — hence its classification as a pulsar. W.50 also seemed to be the wreck of a supernova, and it too contained a radio source, but this radio source showed no sign of pulsar activity.

Later, the British artificial satellite Ariel–5 was sent up to conduct a survey of the sky at X-ray wavelengths. One definite source was found in W.50. The combination of a supernova remnant, a point radio source and an X-ray emitter in the same region seemed to be more than a mere coincidence. Could there be a visible star, or at least some object, which was also connected?

Two English astronomers, David Clark and Paul Murdin, decided to find out. They used the magnificent Anglo-Australian telescope at Siding Spring, in New South Wales, and they soon found the object for which they were seeking. Only later did they realize that it had already been given in a list drawn up by the American astronomers Sanduleak and Stephenson. It was their No.433: hence the designation SS433.

As soon as Clark and Murdin examined the spectrum of SS433, they knew that they had come across something very unusual. There were the expected bright lines due to hydrogen and helium, the most abundant elements in the universe, but there were also other lines which at first defied all attempts at identification. Subsequently it was found that these curious lines moved to and fro across the spectrum — and this really caused a sensation. If a spectral line is shifted towards the long-wave or red end of the band it indicates a velocity of recession, while a shift to the short-wave or blue end indicates approach; this is of course the well-known Doppler effect. SS433 seemed to be 'coming and going' at the same time.

All sorts of theories were put forward, but at last a plausible idea emerged. Evidently we are dealing with a binary system in which the secondary member is a neutron star, produced in an earlier supernova explosion. The neutron star is pulling material away from its companion, and the intense gravitational field heats the infalling material to enormous temperatures, causing intense X-radiation which drives off some of the infalling material in two jets pointing in opposite directions. Now we can see why SS433 appears to be both approaching and receding; one of the jets is moving towards us while the other is moving away. Also, the jets are not fixed in space; they are precessing,

DIAGRAM SHOWING THE SYSTEM OF SS 433

rather in the manner of a gyroscope which is running down and has started to topple. It takes 164 days for the jet to make a complete sweep, and this explains the to-and-fro drifting of the spectral lines. The analogy with the principle of an ordinary garden lawn-sprinkler is obvious.

Further surprises followed. Presumably the material is sent out in jets because the strong magnetic field of the neutron star acts like a nozzle, and radio observations show that the material is moving very rapidly — in fact, at a quarter the velocity of light. As David Clark has pointed out, at this rate a blob of SS433 material could cover the distance from the Earth to the Moon in no more than five seconds. Moreover, the ejection is not smooth; the material comes out in discrete blobs, as the water from a garden sprinkler will do if someone stands on the hose and jumps up and down.

The effects on W.50 itself are quite noticeable. An ordinary supernova remnant would be expected to be more or less spherical, but W.50 has conspicuous 'ears' which must surely be produced by the action of the jets.

At the moment, SS433 is the only known object of its type. Searches for others are going on all the time, but there are reasons for believing that they must be extremely rare. First, we need to start with a binary sytem in the middle of a supernova remnant, but in most cases of supernova outbursts the binary would be disrupted, one member of the pair being driven off into space. Secondly, objects such as SS433 could not persist for very long in such a state. The period of 'lawn-sprinkler' activity may be no more than 100,000 years or so, which is very brief indeed on the cosmical scale.

We usually associate jets with radio galaxies or with quasars, which are beyond our Milky Way system, whereas SS433 is a mere 15,000 light-years away. We must therefore ask whether there is any true analogy. Probably there is, but of course the difference in scale is truly enormous. In SS433 the jets precess in only 164 days, while with quasar jets the corresponding period would certainly be millions of years. Also, a single quasar shines around a hundred times as energetically as an entire normal galaxy. On the other hand, the underlying physical processes may be much the same, and Clark has likened the situation to that of studying elephants compared with ants; examination of ants under a microscope may be able to tell us facts about living organisms which would be equally applicable to elephants — in which case a quasar is an astronomical 'elephant' while SS433 is an 'ant'.

To carry out further researches we need to use not only visual methods, but also radio and X-ray techniques. There is every hope that this will be done in the near future, in which case more objects similar to SS433 will probably be found. For the moment, however, the strange 'lawn-sprinkler' in the Eagle remains unique in our experience.

20 Jupiter's Family

The giant planet Jupiter is a very conspicuous object in the night sky for several months in every year. When it lies close to another planet, the spectacle is truly glorious; thus in mid-December 1986 Jupiter and Mars were neighbours, while in March 1988 Jupiter and Venus passed only two degrees apart. Needless to say, there is no genuine close approach. Mars and Venus are in the foreground, so to speak, and both are relatively small; Venus is not quite so large or massive as the Earth, while Mars has little more than half the Earth's diameter. Jupiter, on the other hand, is big enough to swallow up a thousand bodies the volume of the Earth, and is not solid in the accepted sense of the word. There is a silicate core, but this is overlaid by layers of liquid hydrogen which are in turn overlaid by the dense, gaseous, hydrogen-rich atmosphere which we can see. Obviously, there is no chance of visiting Jupiter. It would be rather difficult to land upon a layer of gas, quite apart from the lethal radiation zones which would quickly kill any astronaut foolish enough to enter them.

Much of our knowledge of Jupiter comes from four space-craft, the Pioneers of the early 1970s and the two Voyagers which by-passed Jupiter in 1979 en route for Saturn. Many problems have been cleared up. Thus the famous Red Spot is not a solid body, but a whirling storm — a phenomenon of Jovian meteorology. We have also learned a great deal about Jupiter's fascinating family of satellites.

Of the sixteen definitely known satellites, twelve are very small, but

JUPITER FROM VOYAGER 1 Two satellites are in transit; the red Io, against the Great Red Spot, and the bright, icy Europa.

the remaining four — Io, Europa, Ganymede and Callisto — are of planetary size. Our own Moon has a diameter of 2,160 miles. Io is slightly larger than the Moon, Europa slightly smaller, and Ganymede and Callisto much larger; indeed, Ganymede, with its diameter of over 3,000 miles, is slightly larger than the planet Mercury, though less dense and less massive.

These four satellites are known as the Galileans, because they were first studied by the great Italian observer Galileo, in 1610 (he was also one of the first to see them telescopically). Galileo realized at once that they were genuine attendants of Jupiter, and this was of great importance, because it showed that there must be more than one centre of motion in the Solar System. At that time the Church was vehemently supporting the theory that all celestial bodies must revolve round the Earth, and Galileo's ideas were regarded as dangerously heretical. It is even said that one leading Churchman refused to look through the telescope, on the grounds that it was bewitched, and that Galileo 'hoped he would have a good view of the satellites on his way to Heaven'!

The Galileans are relatively bright, and but for the glare of Jupiter they would be naked-eye objects. Indeed, a few keen-sighted people can glimpse Ganymede at least without optical aid, and any telescope will show all four. They move round Jupiter at distances ranging from 262,000 miles (Io) out to more than a million miles (Callisto), in periods ranging between 1 day 18 hours for Io up to 16 days $16\frac{1}{2}$ hours for Callisto. Telescopically it is fascinating to watch them from night to night, and to see their eclipses, occultations, transits and shadow transits. Because they move virtually in the plane of Jupiter's equator, they generally appear more or less lined up.

Before the Voyager missions we knew very little about the Galileans themselves, because they appear as tiny disks when seen from Earth, and surface details are well-nigh impossible to make out. However, both the Voyagers sent back close-range views of them, and they have turned out to be very surprising worlds indeed.

Callisto, the outermost, is almost as large as Mercury, but is of low density; there is presumably a rocky core, surrounded by soft ice which is overlaid by the ice-and-rock crust. The entire surface is cratered, and indeed there are so many craters that there seems no room for any more. There are also two large multi-ringed formations, now named Valhalla and Asgard, which are generally thought to have been produced by the impacts of large meteoroids on Callisto early in its history. Callisto today is completely inert; nothing can have happened there for thousands of millions of years. It may well be the first target for any astronauts who manage to reach the Jovian system

100

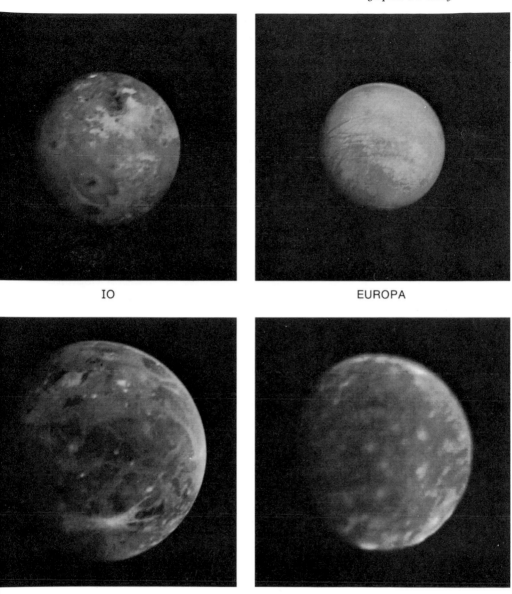

IO EUROPA

GANYMEDE CALLISTO

THE FOUR LARGE GALILEAN SATELLITES The photograph shows them at their correct relative sizes, and the marked differences in their surfaces.

in the future, because at its distance of over a million miles from Jupiter it is well outside the dangerous radiation zone.

Ganymede, slightly larger and more massive than Callisto, is of the same basic type, though there is some evidence of internal activity in its early history. Here too there are icy craters, and one large dark

101

patch which has been appropriately named Galileo Regio. Like Callisto, Ganymede has no detectable atmosphere, and today it must be equally inert.

Though the two largest Galileans have many features in common, the next — Europa, the smallest of the four — is strikingly different. There are practically no craters or vertical relief, and the icy surface is criss-crossed by lines which indicate very shallow depressions. Europa is a map-maker's nightmare, and it is fair to say that one part of it looks very like another.

Why should this be so? Either craters have never been formed, or else they have been obliterated by water or soft ice welling up through cracks in the crust. This has led to the intriguing idea that the silicate interior is hot enough to melt the lower part of the ice-shell, and that below the visible surface there may be an ocean of ordinary water capable of supporting primitive life. Of course, this is highly speculative, and I admit to being sceptical, but I suppose that it cannot be entirely ruled out. Certainly Europa is unlike any other world in the Solar System.

Io, the innermost Galilean, is even more surprising. Voyager showed that the surface is bright red, giving the superficial impression of an Italian pizza. Unlike the other satellites, Io is active, and powerful volcanoes are erupting all the time. Io's volcanoes are not in the least like those of Earth, such as Vesuvius; they are sulphur volcanoes, and the whole surface is sulphur-coated. The general temperature is very low, but some of the volcanic vents are extremely hot, with temperatures of well over 200 degrees Centigrade. Material is hurled high above the ground, and the whole scene must be wildly unstable. There were marked changes in the four-months' interval between the passes of Voyager 1 and Voyager 2; some of the volcanoes became quiescent, while others were even more violent during the second space-craft encounter.

Evidently Io is being 'flexed' by the changing gravitational pull of Jupiter; the orbit is not circular, and the other Galileans also have an effect upon it, though admittedly it is not easy to see why Io is so active and Europa, less than 200,000 miles further out, is not.

The Ionian volcanoes send out ionized atomic particles of sulphur and oxygen which form a doughnut-shaped torus, and the position of Io in its orbit has a marked influence upon the radio emissions sent out by Jupiter itself. Io is connected to Jupiter by a powerful electrical current, and since it is immersed in the radiation zone it is one world which we must be content to examine from a respectful distance.

Jupiter's satellites are in a class of their own, and any owner of a small telescope can follow them from night to night as they move

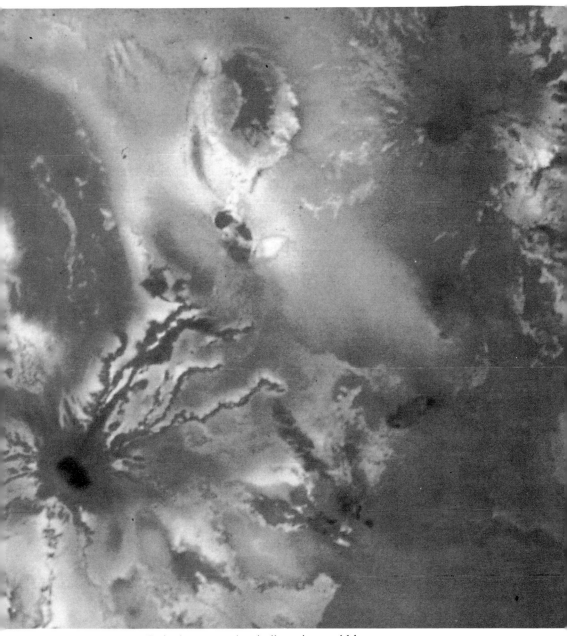

IONIAN VOLCANOES Io is the most volcanically active world known to us.

round their parent planet. As one American astronomer commented soon after the Voyager missions, there is no such thing as an uninteresting Galilean.

103

21 The Sooty Star

Our Sun is a steady, well-behaved star. It does not change much, and has not done so for many millions of years — which is fortunate for us, since even a relatively small change in the Sun's output would have disastrous effects upon life here. Not for several thousands of millions of years in the future will there be any dramatic alteration.

Some stars, however, are much less placid. They brighten and fade quickly, in periods ranging from a few hours to several years, and these changes are genuinely intrinsic. Amateurs carry out very useful work in studying variable stars; there are so many variables in the sky that professional astronomers cannot possibly keep them all under constant watch. Work of this kind is most important, because we are anxious to know how the stars function and how they evolve. Variable stars can provide valuable clues.

They are of different types. Some, such as Betelgeux in Orion, brighten or fade slowly and more or less irregularly. Others, such as Mira in the Whale, reach their maximum brightness every few months (332 days in the case of Mira itself), while the so-called Cepheid variables are as regular as clockwork, and have periods of only a few days. All these stars are pulsating, and are fairly well advanced in their evolutionary story. But there are other variables which are quite unpredictable, so that we never know just what they are going to do next. One of these stars is R Coronae Borealis, about which a remarkable discovery has just been made.

R Coronae lies in the little constellation of the Northern Crown. To find it, first locate the Great Bear, and follow round the 'tail', which leads to a brilliant orange star, Arcturus in Boötes (the Herdsman), which is very prominent during the early months of each year. (It is actually the brightest star in the northern hemisphere of the sky, and is marginally superior to Vega, Capella and Rigel; it is outshone only by the southern Sirius, Canopus and Alpha Centauri.) The little semi-circle of stars marking Corona lies fairly close to Arcturus. The brightest member of the semi-circle, Alphekka, is about the same brightness as the Pole Star.

If you have a pair of binoculars, and look at the Crown, you will certainly see one star inside the 'bowl'; it is just below naked-eye visibility, but any pair of binoculars will show it easily. Probably you will also see another star, on the fringe of naked-eye vision. This is R

A Finder Chart

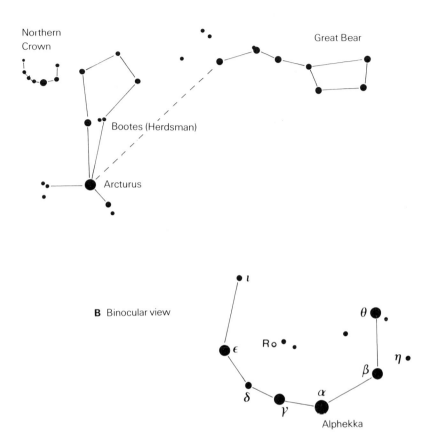

Northern Crown

Great Bear

Bootes (Herdsman)

Arcturus

B Binocular view

ι

θ

Ro

η

ε

β

δ

α

γ

Alphekka

R CORONAE POSITION

Coronae. But sometimes it will be missing; at irregular intervals it fades, taking only a few days to become so dim that powerful telescopes are needed to show it at all. It may remain faint for weeks or even months before it slowly regains its lost lustre. We do not know when these minima are due; R Coronae never tells us.

We are dealing with a very luminous star, well over 5,000 light-years away. We have also found out that in composition it is decidedly unusual. By splitting up its light by means of a spectroscope we can establish what elements are present there, and it is clear that there is much less hydrogen than with normal stars such as the Sun. On the other hand, there is a great deal of carbon, and this gives us a key to the sudden, unpredictable fadings.

What apparently happens is that clouds of carbon particles accumulate in the star's atmosphere. These particles, which are nothing more nor less than soot, form a blanket which shuts in the light coming from below, and the star seems to fade. When the radiation being poured out has become sufficiently powerful, the carbon particles are blown away, and the light can come through once more. It is, then, fair to call R Coronae a 'sooty star'. By plotting its magnitude against time, we can draw up a light-curve; the results given in the light-curve here show a typical minimum which I followed from my own observatory. R Coronae is not unique, but other variables of the same kind are very rare.

In 1983 the Americans launched the IRAS satellite, designed to study long-wavelength radiations from the sky. Cool bodies send out infra-red, and IRAS was remarkably successful. One of the objects surveyed was R Coronae, and three astronomers, Drs Gerry Neugebauer, Fred Gillett and Dana Backman, have found that the star is surrounded by an enormous shell of tenuous material quite unlike anything we have come across before.

The size is truly amazing. The diameter of the R Coronae shell is about 26 light-years — and remember, one light-year is equal to almost six million million miles. If our Sun were the centre of a shell such as this, then many of our nearer stellar neighbours, including Sirius, would lie inside it.

Apparently the shell is made up of thinly-distributed silicate or carbon particles at a constant temperature, which is not hot enough to cause visible light but is detectable at infra-red wavelengths. The American astronomers estimate that it must have been formed over a period of around 125,000 years, ending some 25,000 years ago. But how was it produced, and what is its source of heat?

Frankly, we do not know. It may well be that the material was expelled from R Coronae itself, but the mechanism is obscure. To make matters even more puzzling, it seems that the constant temperature is not due to the central star. There must be some other source; but is it due to an external radiation field, a series of 'hidden' sources, or something quite different? The dust-grains are undoubtedly being heated, because otherwise we would not be able to detect them at all, even in infra-red.

Because the shell of material is so tenuous, it might be thought that it would soon be dispersed by the effects of other stars. This has not happened, presumably because R Coronae lies in an exceptionally 'calm' part of the Galaxy, well away from the central plane.

This, at the moment, is about as far as we can go, and we must await results from a new infra-red satellite. In particular, we want to find out whether other R Coronae type stars have similar shells. One of these stars is SU Tauri, in the Bull, not far from Orion. Generally it is of about the ninth magnitude, so that it can be seen with a small telescope. Like R Coronae, it sometimes fades abruptly, and this happened in early 1988, so that for some months I found that it was extremely difficult to follow with the 15-inch telescope in my observatory.

Undoubtedly R Coronae itself, the brightest member of the class, is of special interest. It has been under observation for many years, and amateurs as well as professionals are always on the alert for sign of a new drop to minimum, but nobody had foreseen that it would turn out to be surrounded by this huge, tenuous, mysteriously-warmed shell. We have much to learn about 'the sooty star'; it guards its secrets well.

22 Flare-up in the Large Cloud

During the spring of 1986 astronomers were paying close attention to Halley's Comet, making its first return to the Sun for seventy-six years. A year later they were preoccupied with an event which was, to be candid, much more important. A supernova flared up in the star-system of the Large Magellanic Cloud, and became visible with the naked eye — the first supernova to attain naked-eye visibility since the year 1604.

The two Clouds of Magellan are among the most familiar features of the far-southern sky. Superficially they look rather like broken-off portions of the Milky Way, but they are in fact independent galaxies, well beyond the limits of our own. The Large Cloud, rather the closer of the two, lies at around 170,000 light-years. It contains objects of all kinds: giant and dwarf stars, clusters, and also the gas-and-dust nebulae in which fresh stars are being born. Unfortunately, it is too far south to be seen from anywhere in Europe.

The Cloud is important because it contains a full selection of astronomical objects at essentially the same distance from us. Many variable stars have been found there, and also many 'new stars' or novae. As we have seen, a nova is a binary system in which the white dwarf component undergoes a sudden, violent outburst which makes it shine with many times its normal brilliancy for a few days, weeks or

months. Novae are not too uncommon, either in our Galaxy or in others, and some of them can become conspicuous; in 1918 a nova in Aquila (the Eagle) temporarily outshone all the stars in the sky apart from Sirius, though it has now become very faint indeed.

A supernova is quite different. A Type I involves the total destruction of a white dwarf, but with a Type II we are dealing with only one star — a supergiant which has used up its essential 'fuel' and is facing disaster.

The initial mass of the pre-supernova star must be at least eight times that of the Sun. After it condenses from its parent nebula, it becomes a Main Sequence star and shines by converting its hydrogen into helium. When there is no more hydrogen available, other reactions take over, and eventually we have a situation in which the star is rather like an onion in structure; there are layers of hydrogen, helium, carbon, oxygen, magnesium and silicon, reckoning inward from the surface. At the core the material is iron. But iron cannot be used as 'fuel', and there comes a moment when the inert core can no longer support the weight of the star's outer layers. In only one second, the core collapses and is turned into an unbelievably dense neutron star; a matchbox full of neutron-star material would weigh well over 2,500 million tons. A shock-wave rebounds from the core, and there is a cataclysmic explosion in which most of the material is blown away to produce a cloud of expanding gas. The remnant — that is to say, the core — will spin rapidly and send out radio waves; it will have become a pulsar. At the climax of the outburst, the temperature may soar to at least 100,000 million degrees, and floods of neutrinos will be emitted.

Because supernovae are so luminous, they can be seen over immense distances; I have already said something about the Type I outburst in the strange galaxy Centaurus A, which was fairly typical. But no supernova in our Galaxy has been seen since the invention of the telescope, and undoubtedly an outburst in the Large Cloud of Magellan was the next best thing. It was discovered on 24 February, when it was already of the fourth magnitude. The discovery was made by Ian Shelton, a Canadian astronomer who was working at the Las Campanas Observatory in Chile, and within hours there was confirmation from New Zealand and Australia. At the Siding Spring Observatory in New South Wales, Robert McNaught had photographed the Large Cloud on the previous night, but had not developed his plates!

PART OF THE LARGE MAGELLANIC CLOUD This photograph includes Supernova 1987a and the Tarantula Nebula. It was taken from the South African Astronomical Observatory on 30 April 1987.

When he did so, there was the supernova, shining as a star of magnitude 6.

How would it behave? It was expected to fade quite quickly, but it did not; after a brief decline it became more brilliant than at the time of its discovery. In early April I made a quick trip to South Africa to have a look at it (after all, I might not have the chance to see another naked-eye supernova for around four hundred years!). It was a sight I will never forget. The magnitude was between 2 and 3 during the week I spent in South Africa, and the star was strongly red; it dominated the whole region, and completely outshone the Cloud itself. It remained visible with the naked eye for some months.

A supernova in a very remote galaxy appears against the blur of the general background, and it is a hopeless task to try to pick out the progenitor — that is to say, the pre-supernova star — on earlier photographs. Not so with this one, which is known officially as SN1987A, because the Large Cloud is so close to us on the cosmical scale. The progenitor was soon identified. It turned out to be a supergiant, catalogued as Sanduleak-69°202, but to everyone's surprise it was not red. It was blue — and the idea of a blue supergiant as a supernova progenitor was completely unexpected. The pre-outburst magnitude was 12.2, and the mass was probably about thirty times that of the Sun.

Initially there were some doubts. Observers studying data from the IUE, or International Ultra-violet Explorer satellite, announced that Sanduleak −69°202 was still there, and so could not have exploded. Some days later it was found that the data had been misinterpreted; the original star had disappeared in its old form. Moreover, bursts of neutrinos were reported at several neutrino observatories, and it may well be that many millions of these strange particles bombarded every square centimetre of the Earth as a direct result of the outburst. Since neutrinos have no electrical charge and (probably) no mass, they are not exactly easy to detect.

The shape of the lightcurve soon showed conclusively that the supernova was of Type II, not Type I. The main surprise — that Sanduleak −69°202 had been blue, not red — also explained why the outburst was under-luminous by supernova standards; had it been conventional, it would have shone in our skies as brightly as Canopus instead of remaining below the second magnitude. A blue supergiant is much smaller than a red one, though equally massive, and there is less surface area to expand. It is this hot, expanding shell which regulates the peak luminosity. The expansion rate was predictably high: of the order of 20,000 miles per second.

At a very early stage it was obvious that SN 1987A was an

exceptional object, and immediate studies were made from all observatories which could see it. At Siding Spring, where the main instrument is the great 158-inch AAT or Anglo-Australian Telescope, it was agreed that 10 per cent of all observing time should be diverted to the supernova, and new equipment was hastily built to cope with it.

Of course, the main research has to be carried out with spectroscopic equipment, and there was a great deal to be done, and not much time to do it. For example, the outburst lights up the material lying between the Large Cloud and ourselves, and from this it was found that there are gas-clouds in space which are much cooler and much more slow-moving than had been anticipated. Another surprise was to learn that the supernova was not round; it seemed to be elongated. There were several possible explanations for this. It could be that the explosion had not been symmetrical; alternatively, the rotating star could have had a magnetic field which would force it to expand more rapidly in one direction than in others; or else the star blew itself apart not as a spherical shell, but in fragments. It could even be that we were seeing light reflected off blobs of gas and dust which were there all the time, but which could not be seen until the supernova illuminated them.

Another intriguing problem was that of what became known as the Mystery Spot — named, I understand, after a popular bar and night-club!

A team from Imperial College, London, went to Siding Spring to carry out what is termed speckle interferometry. This involves taking a number of very short-exposure pictures and then combining them electronically, with the result that the worst effects of the Earth's unsteady atmosphere are 'frozen out'. The London team were doing this when they found a curious blob to one side of the supernova, only a fraction of a second of arc away from it — corresponding to a real distance of about two light-weeks. What exactly was this Mystery Spot?

The first idea was that it could be a cloud of dust and gas illuminated by the outburst. It could not be material from the supernova itself, because it was not close enough to the site of the outburst; to reach it, the material would have had to move at the velocity of light or thereabouts. On the other hand, the Mystery Spot had to be associated with the supernova, because it was much brighter than anything which had been recorded in the area before the outburst.

According to Professor Martin Rees, of Cambridge, the progenitor had a binary companion. As a shell of material spread outward from the explosion-centre, this companion 'got in the way' and blocked some of the shell. Then the companion moved aside, and left what could be regarded as a hole in the shell. Radiation from the supernova poured out through this hole, striking an already-existing interstellar cloud and lighting it up.

What will happen to the supernova in the future remains to be seen. We should be able to keep track of it for years before it is lost in the background glow, and when the débris starts to clear away we ought to find out what the end-product has been. There may be a neutron star, detectable as a pulsar, or else a black hole — a black hole being an area round an old, collapsed star from which not even light can escape, because of the immensely strong pull of gravity. We cannot see a black hole, because it emits no radiation at all, but we can detect it by its effects upon visible objects.

To say that astronomers are fascinated is to put it mildly. They have been presented with an opportunity which may not recur for many centuries, and they are making the most of it. For the first time we may be witnessing the birth of either a pulsar or a black hole.

---

23 Thirtieth Anniversary

The television *Sky at Night* series began in April 1957. In April 1987 we presented a programme to celebrate a thirty-year unbroken run, and it was fascinating to look back and see what had happened during this period. Of course, there were many hitches and amusing episodes, which I will not describe here because I have done so elsewhere;* suffice to say that they were tremendous fun, even if in many cases somewhat hair-raising! So let me now concentrate upon the purely astronomical aspect.

In January 1957 the Space Age had not begun; it did not start until the following October, when Russia's first artificial satellite, Sputnik 1, soared aloft and finally showed that men could indeed reach out towards the planets. Radio astronomy, now a vital part of scientific research, was in its infancy; the great 250-foot 'dish' at Jodrell Bank was not yet in full operation, and in any case it was widely regarded as something of a white elephant. Objects such as quasars, pulsars and black holes were not only unknown, but unsuspected; telescopes still depended upon photography, with only very rudimentary electronics; and the idea of sending astronauts to the Moon, or probes to Halley's Comet, seemed to be looking far into the future.

I suppose the real change in outlook came with Sputnik 1. It was no

*TV Astronomer, (Harrap, 1987).

more than football-sized, and it carried little apart from the transmitter which sent out the famous 'bleep! bleep!' signals never to be forgotten by anyone who heard them (as I did). It was also Sputnik 1 which ended the financial troubles at Jodrell Bank, because no existing instrument elsewhere could act as an effective satellite-tracker, and funds were promptly forthcoming.

It was in 1959 that the Russians went one better by launching the first unmanned probes to the Moon. We had never been able to study the Moon's far side, which is always turned away from us; Luna 3 went on a round trip, and sent back the first pictures of the regions which we had never seen. As expected, they were just as bleak and just as crater-scarred as the familiar areas, though there were fewer of the waterless 'seas'.

At that time the Americans were having trouble with their rockets, but after the initial problems had been solved we came to the years of what has been called 'the race to the Moon'. Whether there ever was a genuine race between America and Russia is a matter for debate, but in any case we had unmanned probes of various kinds, and then, in 1969, the first lunar landing by astronauts. First Neil Armstrong, then Edwin Aldrin stepped out on to the bleak rocks of the lunar Sea of Tranquillity. Up the the mid-1960s there had been fierce arguments about the nature of the Moon's surface, and there were some astronomers who maintained that it would be covered with deep dust-drifts, making landings there more than hazardous. I was no believer in the dust theory, and indeed it proved to be wrong. One fact which was confirmed by the lunar probes was that the Moon is lifeless. Throughout its long history it has been a sterile world, and quarantining of the Apollo astronauts was abandoned as unnecessary after the first two missions.

Mars and Venus were next on the list, and our picture of them today is quite different to what it was in 1957, when both were regarded as possible abodes of life, even if this life was certain to be lowly. And during the 1970s and 1980s unmanned rockets were also able to survey the outer planets Jupiter, Saturn and Uranus, each of which provided its quota of surprises. We did not expect the amazing complexity of Saturn's rings, the dark ring-systems of Jupiter and Uranus, the nitrogen atmosphere of Titan, the varied surface of Miranda, or the red sulphur volcanoes of Io. Recently I looked back at some of the chapters in Volume 1 of *The Sky at Night*, published in 1962. They seem rather quaint nowadays!

Then there was Halley's Comet. In 1957 we expected that if anyone could send a probe to it, we would look to America. In fact the Halley missions came from Europe, Russia and Japan.

113

THE AUTHOR WITH THE ORIGINAL PLANISPHERE BACKGROUND, 1957

To many people it is the space achievements which have dominated astronomy during the past three decades, but this is not a true picture. The most important advances have been in stellar astronomy and cosmology, mainly because of new techniques. A century ago, the photographic plate replaced the human eye for most branches of research; today photography is itself being phased out by electronic devices which are far more sensitive. The telescopes built in earlier years — such as the Palomar 200-inch reflector, which dates back to 1948 — are the same as before, but with the new electronic equipment they are vastly more powerful. We also have great new telescopes, built upon patterns which were quite unfamiliar in 1957.

Let me give you two examples of what I mean. The Palomar 200-inch remained the world's largest telescope for some time, but then the Russians built a 236-inch, which ought to have a much greater light-grasp. To be candid, it has never been a real success, and we have to ask whether we have reached the limit in size so far as single-mirror telescopes are concerned. Probably this is not so, but there are alternatives. One is to make multiple-mirror telescopes. The first of

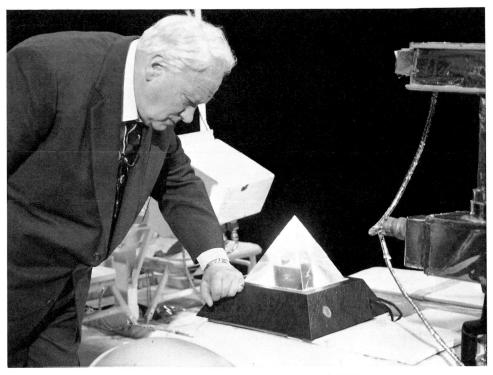

THE THIRTIETH ANNIVERSARY PROGRAMME (SOME YEARS LATER!) This photograph was taken by Mat Irvine.

these was the MMT at the Whipple Observatory atop Mount Hopkins, in Arizona, which uses six 72-inch mirrors working together and bringing the light to the same focal point. The MMT has been a distinct success, and even larger telescopes of the same type are being planned. Telescopes with several 100-inch mirrors are quite practicable. It is easier and cheaper to build several smaller mirrors rather than to cast and figure one colossus.

Another innovation has been the altazimuth mounting. Up to the mid-1970s all large telescopes (and many small ones) were placed on equatorial mountings, so that the up-or-down movements of the targets in the sky looks after itself. With an altazimuth, you have to cope not only with the east-west drift, but also with changes in altitude. Modern-type computers have no problem here, and an altazimuth has many advantages. I doubt if any more giant equatorials will be made; all the latest major telescopes, such as the William Herschel reflector in the Canary Isles, are altazimuth. Note, incidentally, that the first really large altazimuth was the Russian 236-inch, so that in this respect at least it marked a major advance.

115

With these powerful instruments we can probe out to the far reaches of the universe, and study objects so remote that we see them as they used to be when the universe was young. How far we will be able to penetrate in the future remains to be seen.

Then, of course, there is what we may call 'invisible astronomy'. Before 1957 only radio astronomy, and, to a limited extent, infra-red astronomy, had made any real progress. One cause of trouble is the Earth's atmosphere, which blocks out many of the radiations coming from space. Virtually all X-rays are cut out, so that it was not until 1963, with the advent of high-altitude scientific rockets, that X-ray astronomy literally 'got off the ground'. Since then we have had satellites of many kinds, and we can examine the whole range of wavelengths, from the very short to the very long. This in turn has led us on to the discovery of bizarre objects. Quasars were first detected because of their radio emissions, though admittedly we now know that by no means all quasars are strong radio sources. Pulsars, also, were tracked down by radio. The first of them was discovered by accident in 1968, when researchers at Cambridge came across sources which seemed to be 'ticking'. At first it was even thought possible that we were picking up artificial signals from deep space, and it was some days before the LGM or Little Green Men theory was abandoned. Today, it is certain that pulsars are rapidly-spinning neutron stars.

All this has happened during the past three decades.If progress continues at the same rate, who knows what the next three decades will bring forth? We should have new probes to the planets; with luck, bases on the Moon, and permanent space-stations. We will also have the Hubble space telescope, a 94-inch reflector which will operate from 300 miles above ground-level. Freed from the constraints imposed by our air, it will see much farther into space than ever before, and we may even be able to look back to a time only just after the universe, as we know it, was created in the biggest of all bangs.

Computers, of course, provide the key. In 1957 they were elementary; by now they are all-powerful. I am sometimes tempted to ask myself whether the time will come when computers decide that human operators are not really necessary. . . .

Perhaps I can end with a personal note. I know Neil Armstrong, the first man on the Moon; I knew Yuri Gagarin, the first man in space (in fact, both have joined me in *Sky at Night* programmes); I also met Orville Wright, the first airman, who did not die until the late 1940s. This means that Neil Armstrong and Orville Wright could have met, because their lives overlapped. So I have a feeling that I span the ages. I wonder if I have yet to meet the first man who will land on Mars? I would not rule it out.

24 The Air of Other Worlds

We depend entirely upon the Earth's atmosphere. Had not the air been of exactly the right density and composition, life of our type could not have appeared here. What, then, about the other worlds in the Sun's family? Is there any chance that they, too, can have breathable atmospheres?

The presence of an atmosphere depends mainly upon escape velocity, and to a lesser extent upon temperature. Escape velocity is simply the speed which an object must attain if it is to break free without any extra impetus. Thus an object departing from the Earth at 7 miles per second (around 25,000 mph) will not be pulled back, and will move away into space; if the initial velocity is less, then the object will not be able to escape.

Air is made up of millions upon millions of tiny particles, all moving around; if they could work up to escape velocity, they would be lost. Fortunately for us, the gases which make up our air (mainly nitrogen and oxygen) are not as quick-moving as this, and so the Earth has been able to hold on to its atmosphere. Actually, our present air is not the original one. The first atmosphere contained large amounts of the very light gas hydrogen, which leaked away; the air we breathe today was produced later, by volcanic emission from the Earth's interior.

Now consider the Moon, which has only $\frac{1}{81}$ of the Earth's mass. The escape velocity is a mere 1.5 miles per second, and this is not enough for a true atmosphere to be retained. Even if the Moon had a dense atmosphere early in its history — which is by no means certain — it has nothing of the sort now. There is a trace of what may be termed collisionless gas, and traces of helium, neon and argon have been found, but the density is negligible. To all intents and purposes, the Moon is an airless world.

Of the other planets, what about Pluto, whose diameter is 1,520 miles — much less than that of the Moon? One would not anticipate a detectable atmosphere; and yet there is one, made of methane, as has been proved spectroscopically. The density is low, but the atmosphere can be retained because of the intense cold, which slows down the movements of the atoms and molecules. Charon, Pluto's satellite, is only 750 miles across, and no atmosphere has been found, but it has been suggested that Charon's original atmosphere may have been 'stolen' by Pluto. It is even possible that the tenuous methane cloud

117

now envelops both bodies; after all, they are only about 12,000 miles apart.

Mercury is a different proposition. It has a diameter of just over 3,000 miles, but it is as dense as the Earth, and its iron-rich core seems to be larger than the whole body of the Moon. On the other hand, the daytime temperature is very high. A trace of atmosphere has been found, but the main constituent is helium, and it seems that the so-called atmosphere is made up of particles captured from the solar wind by Mercury's magnetic field. In any case, the ground density is no more than a thousand-millionth of a millibar, which is not very much!

We can at once dispose of all the asteroids, only one of which, Ceres, is as much as 500 miles across. The same argument applies to all the satellites of the planets, apart from those which are larger than the Moon; three in Jupiter's system, one in Saturn's and one in Neptune's. About Triton, the senior satellite of Neptune, we know very little, and we must await the results from Voyager 2 in August 1989, but probably there is a thin but appreciable atmosphere. Of the others, Ganymede (escape velocity 1.7 miles per second) and Callisto (1.5 miles per second) appear to be completely devoid of atmosphere. The highly volcanic Io is in a different category, and there is a thin surround of sulphur dioxide; whether it can be classed as a genuine atmosphere is debatable, but the ground pressure is no more than a thousandth of a millibar.

Titan is the oddity. As we have seen, it has a dense nitrogen atmosphere with a surface pressure of one and half times that of air at sea-level; yet the escape velocity, 1.6 miles per second, is not much greater than that of the airless Moon. Of course, the solution is the bitterly cold climate. If Titan were warmer, its atmosphere would quickly leak away as the constituent atoms and molecules speeded up.

In around 5,000 million years from now the Sun will turn into a red giant, and will, sadly, incinerate the Earth, I have often been asked whether Titan will then become habitable. The answer is a regretful 'no'. True, its temperature will rise, but before long it will be left as airless as the Moon is today.

I have already discussed Venus, with its dense, choking atmosphere, which certainly precludes any manned landings there in the foreseeable future. Mars is less unfriendly, but the tenuous atmospheric mantle will be of little direct use to us. On the other hand, there is ample evidence of past river systems, so that Mars must then have had much more air than it has now. We have to ask ourselves why this is so, and whether the atmosphere will ever thicken up again.

Oddly enough, this may not be impossible. The tilt of Mars' axis of rotation alters quite markedly over a long period; when a pole is

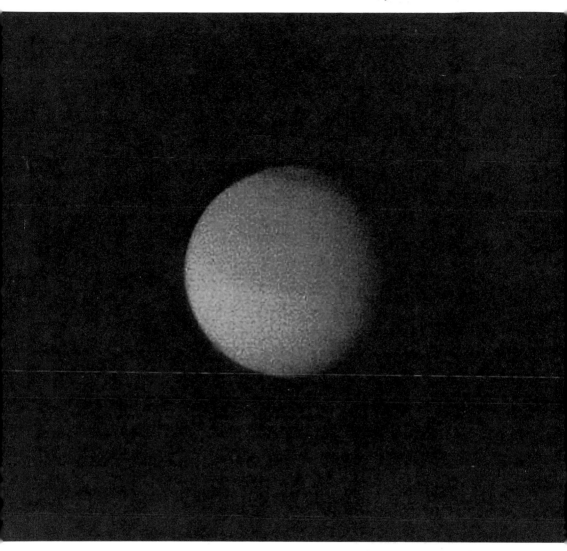

THE CLOUDS COVERING TITAN This photograph was taken from Voyager 1 on 9 November 1980 at a distance of 2.8 million miles.

receiving its maximum amount of warmth from the Sun, the icy cap may disappear, releasing water vapour and making the atmosphere denser for a limited period. Another recent theory is that there are times when the Martian volcanoes become active, sending out vast amounts of gas and vapour. If either of these ideas is correct, we may be seeing Mars at its very worst — but whether it will go through any future 'fertile periods' is very dubious.

119

SATURN AND SIX MOONS A picture from Voyager 1, 17 September 1980, at a distance of 47 million miles. Titan is upper right; Dione and Rhea in the upper left corner (upper, middle and lower respectively; Mimas and Enceladus at the lower right, Mimas being the closer to Saturn.

The surfaces of the giant planets are gaseous, and it is not easy to decide just where the atmosphere ends and the planet proper begins. In any case, we have to concede that apart from Earth, there are no members of the Sun's family with atmospheres suited to our kind of life. On our world, we are lucky. If we did not have the right sort of atmosphere, we would not be here. As Sir Fred Hoyle once commented, we would be somewhere else!

25 The Wandering Moon

Apart from the Sun, which is the most important body in the sky? Most people will probably say 'the Moon', and so far as we are concerned this is true enough, even though the Moon is a very junior member of the Solar System. It illuminates our nights, and is the prime cause of the ocean tides.

Even with the naked eye, you can see the dark patches on the lunar surface — the so-called 'seas' which have never had any water in them. Everyone has heard of the Man in the Moon, the subject of so many legends. But has it ever struck you that the dark patches always appear in virtually the same positions on the Moon's face? For example, the well-marked patch of the Mare Crisium, or Sea of Crises, is always placed near the upper right-hand limb (as seen from the Earth's northern hemisphere); you will never see it anywhere near the centre of the disk, and the other features are similarly fixed. In fact, we always see the same parts of the Moon whenever they are in sunlight.

This means that there is something unusual in the way that the Moon spins, and it is worth looking too at some of the peculiarities of the lunar orbit.

First, it is not strictly correct to say simply that the Moon revolves round the Earth. To be accurate, we must say that the Earth and the Moon move together round their common centre of gravity. However, because the Earth is so much more massive than the Moon, this centre of gravity — or barycentre — lies deep inside the Earth's globe.

Secondly, the Moon's path is not a circle; it is an ellipse. The distance from Earth ranges between 221,460 miles (perigee) and 252,700 miles (apogee), and the effects are quite noticeable; the apparent diameter of the Moon can be as much as $33\frac{1}{2}$ minutes of arc or as little as $29\frac{1}{2}$ minutes of arc. The time taken for the Moon to complete one orbit is 27.3 days; but because both Earth and Moon are moving round the Sun, the interval between one new moon and the next (or one full moon and the next) is 29 days 12 hours 44 minutes. This is a little less than one calendar month, and on occasions February may be bereft of either a new moon or a full moon.

The surface markings appear fixed because the Moon spins once on its axis in exactly the same time that it takes to go once round the Earth: 27.3 days. Therefore, it always keeps the same face towards us. To show what is meant, put down a chair to represent the Earth, and

121

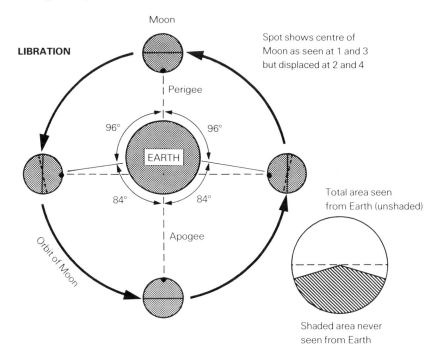

LIBRATION

Moon

Spot shows centre of
Moon as seen at 1 and 3
but displaced at 2 and 4

Perigee

96°   96°

EARTH

84°   84°

Apogee

Orbit of Moon

Total area seen
from Earth (unshaded)

Shaded area never
seen from Earth

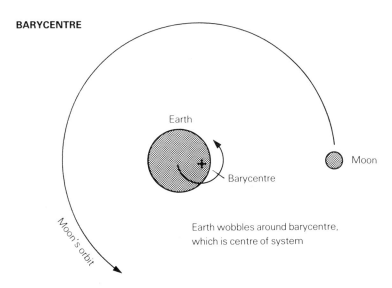

BARYCENTRE

Earth

Barycentre

Moon

Moon's orbit

Earth wobbles around barycentre,
which is centre of system

THE LUNAR SURFACE Photographed by Commander Hatfield, using a 12-in reflector. The Apennines and Alps are shown, with the great Alpine Valley; the Mare Serenitatis is to the left, and you can see the white patch of Linné.

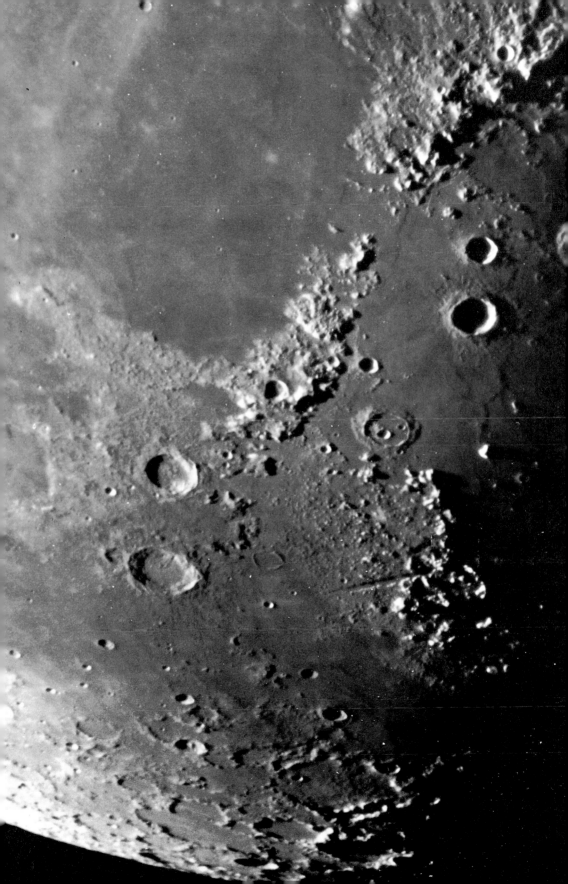

imagine that your face represents the Moon. Walk round the chair, turning so as to keep your face turned chairward all the time. Anyone sitting on the chair will never see the back of your neck, and in the same way we on Earth never see the 'back' of the Moon. Until 1959, when the Russian unmanned probe Luna 3 went on a round trip, we knew nothing definite about the averted hemisphere.

However, there are some complicating factors. Because the Moon's path is elliptical, its orbital velocity varies, whereas it spins on its axis at a constant rate; and periodically the two motions become out of step, so that we can see a little way first beyond one mean limb and then beyond the other. This is the most important of the effects known as librations. There are other librations also, and the result is that we can see a grand total of 59 per cent of the total surface, though of course never more than 50 per cent at any one time.

You may well ask why the Moon has this so-called captured or synchronous rotation. It is not pure coincidence; it is due to tidal forces over the ages. Originally, it is thought, the Earth and the Moon were much closer together than they are now (though the old idea that they used to form one body, and that the Moon broke away, has fallen into disfavour). The Moon was still viscous, and was presumably spinning round fairly quickly, but it had to fight against the pull of the much more massive Earth, which tended to draw out a bulge in the Moon's body. The principle was much the same as that of a bicycle-wheel trying to spin against the influence of restraining brake-shoes. In time the lunar rotation was slowed down until, relative to the Earth, it had stopped altogether. Note, however, that the rotation had not stopped relative to the Sun, so that day and night conditions on the surface are the same in both hemispheres — apart from the fact that to an observer on the far side, the Earth will never be seen at all.

Just as the Earth raised tides on the Moon, so the Moon raised tides on the Earth, though they were less powerful because of the Moon's smaller mass. In the remote past, the Earth rotated more quickly than it does now, as can be demonstrated by studies of fossils. Certain shell-like creatures show daily growth rings which vary in thickness according to the seasons; by examining these rings we can prove that there was a period when there were more than 400 days in a year. The effect continues even now, mainly because of the ocean tides, though the increase is only 0.00000002 of a second per day. Another effect of this tidal interaction is that the Moon is gradually receding from the Earth, but again the effect is so slight, a mere 3 centimetres per year, that you need not be in any hurry to study the Moon before it disappears into the distance!

This state of affairs could not continue indefinitely. Given enough time, the Moon would recede to about 370,000 miles, and the month and the day would be equal at 47 times their present length, after which solar tides would make the Moon draw inward again. However, nothing of the sort will actually happen, because it would take longer than 5,000 million years — and this is the life-expectancy of both Earth and Moon, which are certain to be destroyed when the Sun swells out to become a red giant.

Most people know that the Moon is the main cause of the ocean tides. Actually, tidal phenomena are highly complicated, quite apart from the factors introduced by the irregular outlines of the seas and continents; but basically the Moon tends to heap up water on the moonward side of the Earth, while on the far side the solid globe of the Earth is being 'pulled away' from the water producing a second high tide. The Sun is also a tide-raiser, but solar tides are much weaker than the lunar ones, because the Sun is 400 times further away. When the two bodies are pulling in the same sense, as at new and full moon, we have spring tides, which are relatively strong; when the pulls are opposed, at half-moon, we have the weaker neap tides.

Another effect of the Moon's wanderings was quite evident throughout 1987. The apparent path of the Moon across the sky differs from that of the Sun, because the Moon's orbit is tilted at an additional five degrees, and this is an appreciable amount when you remember that the apparent diameter of the Moon is only about half a degree. During 1987 the difference was at maximum, so that there were times when, as seen from the Earth's northern hemisphere, the crescent moon seemed exceptionally high up and the full moon exceptionally low down.

Finally, what about the famous illusion which makes the low-down full moon look much larger than the high-up full moon? This is well known, and artists usually exaggerate the effect, painting the low moon in a way which makes it look the size of a dinner-plate from close range. Actually it *is* an illusion, and nothing more, but many people find this hard to believe until they make proper measurements.

It is fair to say that the movements of the Moon have taxed the abilities of mathematicians over the centuries. Before the invention of the marine chronometer, the changing position of the Moon against the stars was used as an aid to sea navigation. Today modern computers have solved many of the outstanding problems; and, thanks to radar, we can now measure the distance of the Moon to within an accuracy of one inch.

26 The William Herschel Telescope

La Palma is one of the smaller islands of the Canaries group — do not, please, confuse it with Las Palmas. It is not particularly large; it has only one major town, Santa Cruz de la Palma, and it is not a real tourist resort. Yet it has now achieved world-wide fame, because one of the world's most important observatories has been set up there.

Like all the Canaries, La Palma is volcanic. There is some present-day activity, but the main volcano, Roque de los Muchachos, is extinct — at least, one hopes so. The name means 'the Rock of the Boys', and the Boys themselves are represented by two massive rocks on the summit, at a height of 7,870 feet above sea-level. I am sure there must be a local legend about them, though so far I have never been able to find out what it is.

Astronomers like to build their observatories at high altitudes, so as to be above the densest and most polluted part of the Earth's atmosphere. Nobody can pretend that Britain is a suitable site, and years ago there were plans to shift the main instruments of the Royal Greenwich Observatory away from England to a more favourable climate. La Palma was selected, after a long and exhaustive series of tests, and the new Observatory was inaugurated; the 98-inch Isaac Newton reflector was transferred there, and given a new 100-inch mirror in the process. Other nations, too, were involved, and for some time now there have been several large telescopes in operation. In fact, Los Muchachos is a fully fledged international observatory, though it is situated on a purely Spanish island.

Experiments confirmed that the choice had been a good one. The percentage of clear nights is very high, and the air is normally both calm and transparent, with the cloud-base lying well below the top of the volcano. There was no argument about the location of Britain's newest and largest telescope, the William Herschel Telescope (WHT), named in honour of the great astronomer who discovered the planet Uranus in 1781. Astronomers were unanimous in saying that La Palma was better than anywhere else.

Telescopes are of two main types: refractors and reflectors. A refractor collects its light by means of a lens (a pair of binoculars is simply made up of two small refractors working together), while a reflector uses a mirror. The larger the mirror, the greater the amount of light which can be collected. The largest reflector in the world, at

THE ISAAC NEWTON TELESCOPE AT LA PALMA My picture, taken in 1987. This was easily the largest telescope on La Palma before the William Herschel.

Zelenchukskaya in the USSR, has a 236-inch mirror; next comes the 200-inch at Palomar in California, completed as long ago as 1948. The William Herschel Telescope, with its 165-inch mirror, comes third in order of size, but it is ultra-modern, so that it is really in a class of its own.

As we have seen, photography has long since superseded visual observation for most branches of astronomy, and by now photography is itself being superseded by electronic aids. Zelenchukskaya, Palomar and the rest were designed mainly for photographic work. The William Herschel, on the other hand, was planned and built in the electronic era, so that it has an immediate advantage.

When I first went to La Palma, years ago, the road to the volcano-top was nothing more than a rough track, and I remember that on several occasions my Land-Rover became hopelessly stuck in deep potholes. Today things are very different; a proper road has been built, and was ready by the time that the Observatory was officially opened by the King of Spain. The drive to the summit is fascinating. From the pleasant warmth of the coast, you become aware of a change not only in temperature but also in the scenery; by the time that the Observatory comes into view there is not much left in the way of vegetation, and all the trees lie well below. The first sight of the great domes is impressive, and even though they are so recent they blend in well with the overall picture. One tends to think that they have always been there.

The Isaac Newton Telescope, of course, has its own dome, and so have the other instruments — for example, the 40-inch Kapteyn telescope, named after the famous Dutch astronomer, and the 'solar tower' which is Swedish, and was actually the first fully operative instrument on Los Muchachos. But it is the William Herschel dome which really dominates the scene.

Go inside, and you will have your first view of the telescope itself, which — like all modern reflectors — has a skeleton rather than a solid tube; after all, the only function of a mounting is to keep all the optics in the correct positions. It has been said that the telescope seems to fill most of the dome, and this is by deliberate intent, because the mounting is of the latest altazimuth type rather than the conventional equatorial. I have mentioned this earlier, but I hope you will forgive me for repeating it, because so far as the WHT is concerned it is all-important.

As the Earth rotates, all the bodies in the sky seem to move round from east to west. A telescope has therefore to be guided so as to compensate for this movement; otherwise the target object would move quickly out of the field of view, as every amateur astronomer knows. With an equatorial mounting the telescope is set upon an axis, which means that the up-or-down motion is automatic; only one driving mechanism is needed. With an altazimuth, the telescope can move in any direction, so that there must be two drives — one in *alti*tude and one in *azimuth* (east to west). Obviously, this is a complication, and until recently it was too difficult to be attempted, even though it makes most of the other problems much easier. Today, with our sophisticated computers, there is no such difficulty. It was the Russian 236-inch which pioneered the computerized altazimuth mounting for giant telescopes, and the WHT has been built upon the same pattern. It means that the whole instrument is much less cumbersome than it would otherwise be, and it can fit well into a more compact dome. Neither are there any voluminous spaces which can fill up with hot air and cause unacceptable currents. The inside temperature is strictly controlled, with cavity walls, and screens protecting the telescope from the daytime sun.

There is little doubt that the 165-inch mirror is as good as any ever made, and its accuracy is a purely British triumph; the main work was carried out by the late David Brown in the Grubb Parsons optical workshop in Newcastle (sadly, the last mirror which will be made there, since Grubb Parsons have now closed their optical department). The focal length of the mirror is 34.4 feet, or, in scientific terms, f/2.5. It has to be made as reflective as possible by means of an ultra-thin layer of aluminium, and this involves a huge aluminizing tank which is in use for the mirror for a total of one hour per year, but is absolutely vital to the telescope's performance.

There are various optical systems, and the change-over from one system to another is a quick, easy process. Moreover, the time is long past when an astronomer had to spend cold, weary hours at the eye-end of his telescope, checking to make sure that the target was still firmly in the correct position in the field of view. Today the observer is not in the main dome at all during the observing run; he may be in a comfortable control-room, watching a television screen, or he may not even be in the observatory at all. The WHT can be operated by remote control, and it is quite possible that the observer using it may be as far away as Britain.

The WHT is not designed to study comparatively nearby bodies such as the Moon and planets. Its main rôle is to investigate the very faint, far-away objects which are much too dim to be made out at all with smaller telescopes. In particular, it is ideal for cosmological research. For this kind of work we need the best available equipment operating under the best possible conditions, and the WHT is a major step forward. Of course, it will be surpassed in size before very long, but for the moment it remains the greatest of all the telescopes of the new electronic age.

Within a year of its inauguration the WHT had shown its worth. It is every bit as good as had been hoped, and it has a brilliant future. We will learn much from the great new telescope on the summit of 'the Rock of the Boys'.

27   The Other Side of Hawaii

To most people Hawaii conjures up a picture of sun-baked beaches, palm-trees and Bikinis. This is true enough; if you want an exotic holiday, Hawaii is the ideal place. The main town — Honolulu, on the island of Oahu — .is to all intents and purposes a large, normal American city, though the scenery all round is magnificent. Big Island is different. Here the principal town, Hilo, is comparatively small, and the total island population is much less than that of Oahu. Big Island is also the only part of Hawaii which is highly active volcanically.

A volcano is produced over a 'hot spot' in the Earth's mantle, deep below ground-level. However, the Earth's crust shifts slowly over the mantle, so that in the course of time the volcano will move away from the hot spot and will cease to erupt. On Big Island there are two particularly large and massive shield volcanoes, Mauna Kea and Mauna Loa, both of which are around 14,000 feet high. Mauna Kea has left the hot spot, and has become dormant; it has not erupted for at least four thousand years, and will probably never do so again. Mauna Loa, on the other hand, is extremely active. It can erupt with tremendous violence, and once, during the last century, lava from it even rolled as far as the outskirts of Hilo — where, according to legend, it was stopped only by the intervention of a powerful witch-doctor who had been hastily summoned to take charge!

Adjoining Mauna Loa is Halemaumau, one of the most continuously active volcanoes in the world. The great 'fire pit' is spectacular at any time; when it erupts it sends fiery fountains high into the air, producing a superb display of natural pyrotechnics. Unfortunately, I have never seen this myself. I have been there several times, but Halemaumau has never had the courtesy to perform in my presence. Apparently, though, there was every reason for the Hawaiians of yesteryear to associate it with the formidable fire-goddess Pele, who was not noted for her tolerance or good nature.

Leaving Hilo, you can drive up the long Saddle Road, which will eventually take you to the other part of the island. Before long the houses are left behind, and the road becomes quite steep; you leave the lush vegetation, and soon there is nothing left but very scrubby plants which somehow eke out an existence upon the black lava. The upper part of Saddle Road is desolate indeed, and on occasions it can be blocked by the lava from Mauna Loa, though this does not happen often. Saddle Road winds its way between Mauna Loa and Mauna Kea, so that the two huge shield volcanoes — the largest in the world — are to either side. You will be aware of a marked drop in temperature, and the air becomes decidedly thinner than it is in Hilo.

Eventually the road forks. Saddle Road itself starts to slope downward on its way to the far side of Hawaii. If you turn right, you will continue to climb. This is the way to the Mauna Kea Observatory, one of the world's greatest astronomical centres and also the loftiest — with the sole exception of the high-altitude observatory at Boulder in Colorado. The scene becomes even wilder, with little plant life left, and at last the summit of Mauna Kea comes into full view.

Working at these heights can be a hazard. At 14,000 feet your intake of oxygen is only 39 per cent of its value at sea-level, and some people are badly affected by it; I have known a couple of young, strong athletes who could not cope with it at all. To 'break yourself in', the recommended advice is to stop for at least twenty-four hours at Hale Pohaku, the base station for the Observatory, which is at just below 10,000 feet, and where the air is still dense enough to make most visitors feel fairly comfortable, though violent exercise is not to be recommended. (Hale Pohaku has a pool table in its main recreation room, but table tennis is absent.) I have never had any altitude trouble, because I used to fly around in open cockpits during the war and I know just how to breathe, but it is wise to be careful even at Hale Pohaku, and certainly on the summit itself.

The road from Hale Pohaku to the Observatory is labelled 'Dangerous. Travel at your own risk', and four-wheeled drives are more or less essential. To climb the last four thousand feet takes only about twenty minutes, and the road is by no means so bad as might be expected, though it is distinctly rough in places. You pass by craters, lava-flows and volcanic hills; everything is black, apart from snowy patches which seldom disappear even in summer. Then, quite suddenly, you turn a corner in the road and come to the Observatory itself.

It was the brainchild of Gerard Kuiper, a Dutch astronomer who spent much of his life in America and was one of the main architects of the plans to send space-probes to the planets (it was sad that he died too soon to see the most important results). Kuiper was concerned with setting up equipment above the densest part of the Earth's air, partly because the atmosphere is dirty and unsteady, but mainly because it blocks out many of the most interesting radiations coming from space. Remember, visible light makes up only a very small part of the total range of wavelengths, or electromagnetic spectrum. If the radiation has wavelength longer than that of red light, we cannot see it; we come first to infra-red, then to microwaves, and then to the much longer radio waves. From sea-level we can study little of the infra-red or microwave region, and only some of the radio waves, so that the only answer is to go aloft. Satellites and space-stations are invaluable, of course, but much can also be done from 14,000 feet, and Kuiper set his sights firmly on Mauna Kea.

He met with a great deal of opposition, but he persisted, and at last he had his way. Building a major observatory under such conditions is far from easy, to put it mildly, but it can be done, and by now there are several of the world's best telescopes in operation there. Occasional ground tremors occur, and now and then clouds of ash from Mauna Loa interfere with the seeing conditions, but on the whole Mauna Kea is co-operative. From its summit, Mauna Loa is a magnificent spectacle in the distance.

One of the most important telescopes is UKIRT, the United Kingdom Infra-red Telescope. This looks very like an ordinary reflecting telescope, and can indeed be used as such, but it was designed to collect the longer-wavelength radiations, and it has been remarkably successful. A new type of camera, developed by astronomers at Edinburgh, can even be used to take direct infra-red pictures, which is a major step forward.

Infra-red can penetrate interstellar dust which blocks out ordinary light completely, so that we can study regions which would otherwise be permanently hidden. Consider the Orion Nebula, a huge dust-and-gas cloud over 1,000 light-years away. (It is easily visible with the naked eye, below the three stars of the Hunter's Belt.) Deep inside it, new stars are being born from the nebular material, and there are some strong infra-red sources; one of these, known as the Becklin–Neugebauer Object after its discoverers, is now thought to be a vast, immensely powerful star whose light can never escape from the nebula. UKIRT has been in action for some years now, and has been responsible for many discoveries. It could never operate efficiently from sea-level; the Earth's atmosphere would make sure of that.

There are other telescopes, too, operated by various nations. The latest addition is again British (with Dutch collaboration), and has been named the JCMT or James Clerk Maxwell Telescope in honour of the last-century Scottish scientist. It looks quite different from

MAUNA KEA, 1987 The observatory of the James Clerk Maxwell Telescope is in the left foreground.

UKIRT, because it deals with radiations of wavelength 0.3 to 3 millimetres; it should really be classified as a radio telescope, and it does not produce a visible image. It is ideal for studies of relatively cool objects, and it is a pioneering instrument, because up to now comparatively little work has been carried out at wavelengths of a millimetre or so.

The collecting 'dish' is 15 metres in diameter, and is amazingly accurate, because the error in the curve cannot be more than the thickness of a sheet of writing-paper. It is in use for twenty-four hours a day, and, rather surprisingly, it is generally used with a protective membrane covering the whole of the viewing slit in the rotatable dome. This is to guard against overheating from the Sun, and also against wind force; believe me, the winds on top of Mauna Kea can be violent, and are capable of blowing at up to 200 miles per hour. The membrane was actually constructed by sail-makers on the Isle of Wight, and is

133

coated with a very thin layer of Teflon, which cuts out very little of the microwave radiation — though it can be rolled back during observations of very weak sources. The JCMT has already been responsible for detecting the first microwaves from a quasar, 3C–273, and it is also being used for studies of the famous microwave background, which consists of weak radiation coming in from all directions all the time. This background has been known for over twenty years, and was really responsible for the rejection of the once popular 'steady-state' theory, according to which the universe has always existed and will exist for ever. The feeble, persistent radiation may well be the remnant of the original big bang which took place some 15,000 million years ago, when the universe, in the form which we know, came into existence.

A second microwave telescope, with a 10-metre dish, has been set up by scientists of the California Institute of Technology. It is close to the JCMT, and plans are already being made to use the two instruments together.

Gerard Kuiper's dream has become reality, and as a site Mauna Kea is unrivalled, but the high altitude does present problems; at 14,000 feet you cannot think as quickly or as clearly as you can at sea-level, or even at Hale Pohaku. Nobody actually sleeps at the Observatory. At the end of an observing run, the astronomers come down to the Base. There have been occasions when observers have been marooned on the summit by sudden bad weather, but the road is now much better than it used to be, so that the chances of being isolated are less.

New telescopes are being prepared. One of these — nearly ready — is the Keck Telescope, so named because seventy million out of the total cost of ninety million dollars was donated by a Mr Keck. There will be a 10-metre mirror for use at visible wavelengths; the mirror will be a new innovation, since it will be made up of segments which are fitted together to form the correct optical curve. Other instruments will follow. A few years ago the summit of the old volcano was deserted and lonely, visited by few people; today it is a hive of activity both day and night. I wonder what the fire-goddess Pele would have said?

28   The Hunt for Planet Ten

Is there a tenth planet in the Solar System, moving far beyond the orbits of Neptune or Pluto? Many astronomers believe that there is. I certainly do. But belief is one thing, proof another; and we have to admit that as yet we have no idea where Planet Ten is likely to be.

To put the situation in proper perspective, let us look back to see how the known Solar System has been extended in near-modern times.

The five naked-eye planets must have been known before the dawn of human history. Mercury and Venus are closer to the Sun than we are; Mars, Jupiter and Saturn are further away. Together with the Sun and Moon, this made a grand total of seven — and seven was the mystical number of the ancients. Nothing could have been more satisfactory.

Then, in 1781, a Hanoverian-born organist named William Herschel, observing from the city of Bath with a home-made telescope, caused an astronomical sensation by discovering a new planet, much more remote than Saturn. It is true that Herschel initially mistook it for a comet, but it did not take long for its true nature to be recognized, and Herschel became world-famous; subsequently, of course, he became probably the most skilful observer in history. After some discussion, the planet was named Uranus. Not until 1986, with the pass of the space-probe Voyager 2, did we know a great deal about it, but at least its movements were carefully studied.

Uranus is a giant world, just over 30,000 miles in diameter; it is just visible with the naked eye if you know where to look for it. It takes 84 years to go once round the Sun, at a mean distance of 1,783,000,000 miles (as against 886,000,000 miles for Saturn). It has a gaseous surface, though in composition it differs markedly from Jupiter or Saturn.

When a planet is found, the mathematicians set to work to decide how it should move. However, Uranus refused to behave. It persistently strayed away from its predicted path, so that clearly some unknown force was acting upon it. Two mathematicians, John Couch Adams in England and Urbain Le Verrier in France, independently decided that this force must be an unknown planet; they worked out where it should be, and in 1846 it was found by Johann Galle and Heinrich D'Arrest, at the Berlin Observatory, very close to the expected position. It was named Neptune, and proved to be slightly smaller but appreciably more massive than Uranus.

Despite their accurate predictions, both Adams and Le Verrier had the orbit wrong. They had expected their planet to be much farther away from the Sun than Neptune actually is. Adams gave a period of 227 years and Le Verrier one of 217 years; the true period of Neptune is only 165 years. I have no doubt that both mathematicians were influenced by Bode's Law, an alleged mathematical relationship, linking the distances of the various planets from the Sun. It works reasonably well out to Uranus, but it breaks down completely for Neptune, and personally I regard it as pure coincidence — a sort of 'take-away-the-number-you-first-thought-of' exercise. However, it was taken very seriously by many last-century astronomers, and it still has some supporters.

Once again the Solar System was regarded as complete, but nagging doubts remained; there were still slight discrepancies between the predicted and the observed motions of Neptune and (particularly) Uranus. The problem was taken up by Percival Lowell, founder of the major observatory at Flagstaff in Arizona. Sadly, Lowell is best remembered today for his wild theories about intelligent canal-building Martians; but he was an excellent mathematician, and by the early years of our own century he had worked out a position for yet another world — known variously as Planet Ten or Planet X.

He searched, but in vain, and when he died in 1916 the planet had refused to show itself. For a while nothing more was done, though independent calculations by W.H. Pickering, also an American, led to much the same result. Eventually astronomers at the Flagstaff Observatory decided to try again. A young amateur named Clyde Tombaugh was called in, given the use of a fine 13-inch refractor obtained specially for the purpose, and invited to set to work. In 1930, only about a year after he had started, he located the planet we now call Pluto.

Tombaugh, of course, was searching photographically, and the conditions were quite different from those experienced by Galle and D'Arrest. They had been looking in a specific position, and discovered Neptune on the first night of their hunt, whereas Tombaugh was working his way along the ecliptic and had much less definite information to guide him. All the same, Pluto turned up not so very far from the position given by Lowell; and later it was found that the planet had earlier been recorded on one of the Flagstaff photographs, and also by Milton Humason, at the Mount Wilson Observatory, who had carried out a brief search on the basis of Pickering's calculations. Pluto had been missed because it was unexpectedly faint. Moreover, Lowell's calculated orbit was wrong; he had given a period of 282

THE DISCOVERY OF PLUTO This is the discovery picture taken by Clyde
Tombaugh in 1930. The over-exposed star is Delta Geminorum.

years, whereas Pluto's real period is 248 years.

It was a tremendous triumph, and all credit must be given to Clyde
Tombaugh,* but almost at once serious doubts began to creep in.
Pluto was an enigma. Its orbit was unusual; it was much more
eccentric than those of the other planets, and for part of the time it is
actually closer in than Neptune. This is the case at the present
moment. The next perihelion, or closest approach to the Sun, is due in
1989, and not until 1999 will Pluto regain its title of 'the outermost
planet'. There is no danger of a collision with Neptune, because
Pluto's orbit is tilted at the relatively high angle of 17 degrees.

But the real problem was Pluto's small size and low mass. Initially it
was thought to be larger than the Earth, but as better measurements
were made this was found to be wrong. Finally, in 1977, it was found
that Pluto has a satellite — Charon — and it became possible to get a
really reliable estimate. We now know that Pluto is a mere 1,520 miles
in diameter, much smaller than the Moon, while Charon is only 750
miles across. Both are presumably made up of a mixture of rock and
ice. By planetary standards they are of negligible mass, and they could
not possibly exert any detectable influence upon the movements of

*I first heard of the discovery of Pluto during a science lesson at my prep school.
Little did I think that fifty years later Clyde Tombaugh would do me the honour
of inviting me to collaborate with him in writing a book about it!

giants such as Uranus and Neptune. One might as well try to divert a charging elephant by throwing a baked bean at it. Either Lowell's reasonably correct prediction had been sheer luck, or else the real Planet Ten awaits discovery.

I cannot accept the luck theory. It would be too much of a coincidence, even allowing for the fact that Lowell was less accurate than Adams and Le Verrier had been for Neptune. One can juggle with the figures, but it is worth remembering that Adams and Le Verrier had been very close to the mark *even though they had the wrong orbit and the wrong mass*. So it looks as if Planet Ten must exist.

Locating it is bound to be a major problem, because it must be very faint indeed, and we cannot hope for success unless we have at least a rough idea of its position. Once again the mathematicians have set to work.

Some years ago Dr Brady, in America, turned to our old friend Halley's Comet, which has a period of 76 years and spends part of its time well outside the orbits of Neptune and Pluto. Brady even gave a position for his hypothetical planet, but nothing was found, and it now seems that his calculations were unsound. More recent work by Dr John Anderson, at NASA, seems to be much more promising.

Anderson has examined the observed positions of Uranus and Neptune over the years, and has come up with some surprising conclusions. It seems that before 1910 or thereabouts there were slight but definite discrepancies between theory and observation, but these discrepancies have now disappeared. The intriguing possibility is that Planet Ten has a very eccentric orbit, and is also highly inclined, so that its path makes almost a right angle with those of the other planets of the Solar System. The mean distance from the Sun is given as around 7,400 million miles, with a period of between 700 and 1,000 years and a mass about five times that of the Earth. If the planet were near perihelion at some time before 1910, and has now receded so far that its effects have become too slight to be detected, we can account for its apparent vanishing act. At its maximum distance from the Sun, the planet could even pass through the Oort Cloud of comets, perturbing some of them and sending them inward.

Here we may enlist the aid of two space-craft, Pioneers 10 and 11, which by-passed Jupiter in 1973 and 1974 respectively. Pioneer 10 simply went on out of the Solar System, but Pioneer 11 was swung back to a rendezvous with Saturn, in 1979, before it too began a never-ending journey away from the Sun. Both are still sending back signals, and we know exactly where they are; we may hope to follow them for some time yet. It is conceivable that Planet Ten might perturb them, and thence betray its position. Remember too that Pioneers 10 and 11

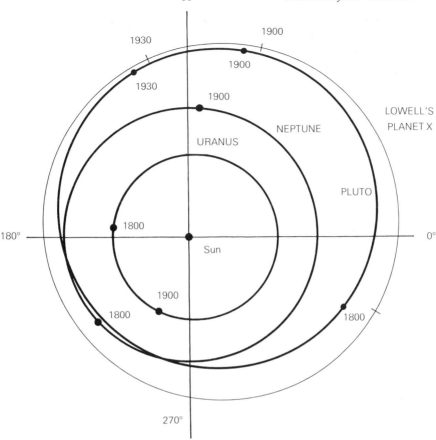

ORBIT OF URANUS, NEPTUNE, PLUTO AND LOWELL'S HYPOTHETICAL PLANET X

are leaving the Solar System in opposite directions, which gives us a double chance. Up to now no perturbations have been found, but one never knows what may happen in the future.

Of course, all this is highly speculative, but it is based upon something more substantial than guesswork, because there does not seem much doubt that there have been tiny irregularities in the movements of Uranus and Neptune in past years, and it is not easy to find any alternative explanation. Whether Anderson's conclusions are valid remains to be seen.

If Planet Ten really exists, it must be lonely beyond belief. From it, the Sun will appear as nothing more than a brilliant point of light; the temperature will be so low that any atmosphere would freeze; the inner planets could never be seen, and beyond there would be a void of millions of millions of miles before coming to the nearest star. But to find it would be a triumph indeed, and most astronomers feel that sooner or later the elusive Planet Ten will make its presence known.

29 Jodrell: The Listening Bank

In October 1987 I was privileged to attend an important scientific ceremony. It was held at Jodrell Bank, near Macclesfield in Cheshire, home of the great 250-foot radio telescope. Jodrell Bank was exactly thirty years old in its present form, and to mark the occasion the 250-foot was renamed the Lovell Telescope.

The telescope itself is immensely impressive; the total weight is some two thousand tons. It is a mechanical marvel, but without the skill, energy and courage of one man, Professor Sir Bernard Lovell, it would never have been built, and the science of radio astronomy would not be where it is today.

Yet its early days were distinctly chequered. As Sir Bernard told me, research developed along unexpected lines. 'The experiment I originally intended to carry out has never been done,' he said. 'When I came here we immediately discovered the whole new field of what became known, some years later, as radio astronomy. First, in 1947–8, we built a big wire paraboloid, 280 feet in diameter, which could only look toward the zenith. Then I managed to find an engineer, Sir Charles Husband, who said that it was possible to build a device of that size which was steerable. After many traumatic experiences this developed into the instrument we now see. Russia's Sputnik came along at a good time for us. A few days before it was launched I said to one of my colleagues that we needed a miracle to save us, because we were in such disfavour and in so much debt. That miracle did come, on 4 October 1957, from behind the Iron Curtain.

'When the telescope was built it was really an act of vision, because there was not much idea of what astronomy we would actually be doing. In those days, the work we are now carrying out would have been practically unrecognizable. It is exciting to have seen so many new things, new concepts, during those thirty years. At the beginning of the period we didn't know about pulsars, we didn't know about quasars, we certainly didn't know about megamasers and those phenomena which are really at the heart of modern astronomy.

'We had no idea at all of the way in which things would develop. All we were really doing was to pick up radio waves from the sky; we didn't know where they came from, and we didn't realize that we would become astronomers. In fact, we didn't call ourselves astronomers at all.

'Some people still think that radio astronomy and optical astronomy are two entirely different subjects, not connected in any way, but of course this is quite wrong. There is a wide spectrum, all the way from radio through to optical waves, X-rays, gamma-rays and so forth, and we play our part in it. After all, radio astronomers use optical telescopes as well. If we have an object such as the Crab Nebula with its pulsar, which emits over the entire spectrum, we may work on it here from its radio waves, but then we go and use optical telescopes to look at the other parts of its spectrum. In the same way, optical astronomers come here. It's all complementary.'

When Sir Bernard retired as Director he was succeeded by Sir Francis Graham-Smith, the Astronomer Royal, who has carried on the work, and has also established a major exhibition, complete with planetarium. There is also Merlin — an acronym for Multi-element Radio-Linked Interferometer — of which Jodrell Bank is the central pivot. To quote Dr Tom Muxlow, one of the main researchers:

'Merlin is a collection of telescopes around England. There is one telescope here at Jodrell, and six others scattered around, joined to Jodrell by radio links which are used to send back their data. If we want to make radio images with resolution comparable with that of optical telescopes, we need a physically large instrument, because radio waves are of much longer wavelength than light-waves. Most of our theoretical telescope doesn't exist! It is clearly impractical to fill England with telescopes, and therefore we have only a few 'sample strips'. This has the effect of producing a very poor raw image, and to improve its quality we have to use a powerful computer. Objects being studied include exploding galaxies, quasars and much else.'

What other discoveries have been made from Jodrell in the recent past? Dr Andrew Lyne has discovered the first ultra-rapid 'millisecond pulsar' in a globular cluster; Dr Jim Cohen has been studying cosmic megamasers; Professor Rod Davies has concentrated upon hydrogen abundances in nearby galaxies; Dr Richard Davies has been examining fluctuations in the microwave background; and so on. All this work is of tremendous importance, and during the programme I was able to discuss some of the programmes in detail.

Consider masers. 'Maser' stands for Microwave Amplification by the Stimulated Emission of Radiation, and a cosmic maser works rather as follows. We begin with an OH molecule (O standing for oxygen, H for hydrogen). When OH molecules are bombarded by infra-red radiation, their spin is increased, and therefore they end up with more energy than they would otherwise have. This does not last. They cascade back to their 'ground state', so to speak, but not completely; a certain excess is left, and this is cumulative, so that when enough OH molecules have been affected in this way the final result is tremendous emission of microwave radiation. To quote Dr Jim Cohen:

'A megamaser is a new type of maser, millions of times more powerful than those we know about in our Galaxy. It is as if you could take the energy output from about a thousand stars like our Sun and convert it to a single narrow microwave signal. The signals are produced by those hydrogen molecules which are in molecular clouds — that is to say, clouds of dust and gas in the centres of certain types of active galaxies. The nature of these galaxies became apparent only when results came in from the IRAS satellite, and it turned out that the megamaser signals were being produced by galaxies which are extremely powerful infra-red emitters. The energy sources themselves are hidden by the dense clouds of molecules, together with gas and dust.

'The source of this amazing power is uncertain, and at the moment there are two competing theories. One states that there is a single very powerful source, perhaps similar to the nucleus of a Seyfert galaxy;* in this case we would have a scaled-up version of the maser situation in our own Galaxy, where we have cocoons of gas and dust around young stars. Alternatively, we may be dealing with a star-burst, in which millions of stars are all born at essentially the same moment in a very confined space in the nucleus of the galaxy. At present, the evidence is marginally in favour of the single energy-source. The reason for this is that when we look carefully at the spectra we find that the gas in the centre is moving very rapidly, at 1,000 kilometres per second or so, and this is the type of motion you would expect to find in the nucleus of a Seyfert galaxy. It is not the sort of motion you would expect to find in the case of a star-burst situation.'

As well as the Lovell Telescope, there is another large 'dish' at Jodrell. This is the Mark 2 telescope, which Dr Richard Davies has been using together with the Lovell in order to study fluctuations in the microwave background. 'The real mystery is why there are galaxies at all; why are the stars in the universe not spread about evenly? Why do we have large 'islands' of stars and then large gaps until we come to the next galaxies? It seems that the original fireball must have been subject to fluctuations, with some regions which were bright and dense and others which were fainter and more rarefied. In time, the bright regions would evolve into galaxies. Therefore, we are trying to measure these fluctuations, but they are proving extremely difficult to detect. Using equipment both here and at Mount Teide, in Tenerife, we are starting to trace them, but they are surprisingly weak. This means that there is a real problem as to how the galaxies came into

*A Seyfert galaxy is a galaxy with a bright, condensed centre or nucleus and only weak spiral arms. All of them seem to be extremely active.

'57–87' PAINTED ON THE BOWL OF THE 250-FOOT DURING THE RENAMING
CEREMONY

existence, and we are starting to think that conventional material
(baryonic material: that is to say, protons and neutrons) cannot be
responsible for most of the mass of the universe. We refer to the
hidden mass as 'cold dark matter'. Neutrinos may be an answer, but
certainly we have to consider some form of physics which we do not yet
understand.

From all this it is very clear that in this sort of research Jodrell Bank
leads the world. I was present when the telescope was lowered during
the ceremony, with '57–87' painted on the inside of the dish. It was a
dramatic moment, and Sir Bernard summed it up aptly: 'This could
not have been foreseen thirty years ago. Now I think we may be
confident that the telescope will survive for many more years, and will
do marvellous work in the future.'

143

30 What Telescope should I buy?

Over the years I have had countless letters from people asking for my advice about telescopes. The theme is always the same. 'I have become interested in astronomy, and I need a telescope. What type ought I to buy, and how much will it cost?'

Letters such as this are not easy to answer, because the unpalatable fact is that astronomical telescopes are either good or cheap — not both. When I was aged ten, I managed to buy an excellent 3-inch refractor for the princely sum of £7 10s, and I still use it. But that was in 1933; today the same telescope would cost nearer £500! Also, it used to be possible to buy telescopes second-hand. Today a really good, cheap, second-hand telescope is about as common as a great auk.

To recapitulate: there are two types of astronomical telescopes. In a refractor, the light from the object to be studied is collected by a lens (object-glass) and brought to focus, where the image is magnified by a second lens known as the eyepiece. In a reflector, the light is collected by a curved mirror. In the Newtonian form, the light is then sent back up the tube on to a second, smaller mirror and is directed into the side of the tube, where an image is formed and magnified by an eyepiece as before. The upper end of a Newtonian is open, and the 'tube' may well be a skeleton construction.

First, let us consider the refractor. The main emphasis must be upon aperture: that is to say, the diameter of the object-glass, which is solely responsible for collecting the light. All the actual magnification is done by the eyepiece, but only a certain amount of light is available, so that the maximum magnification is limited. In my opinion the greatest useful power is 50 per inch of aperture. That is to say, if you have a 3-inch refractor (object-glass 3 inches in diameter) you can use a power of up to $3 \times 50 = 150$, so that the object will appear 150 times larger than it would if seen with the naked eye. In theory, you can use a higher power — say $\times 300$ — but the image will be so faint that it will be quite useless.

With a Newtonian reflector, the aperture is the diameter of the main mirror. Again, the rule of thumb is 50 per inch of aperture, though a reflector is less efficient than a refractor of the same aperture simply because the light from the target object has to undergo several reflections. True, there are various other patterns of reflectors using both mirrors and lenses, and these are excellent, but unfortunately they are also very expensive. My own view (and others may not agree!) is that there is not much difference in efficiency between a good 3-inch refractor and a good 6-inch Newtonian.

I consider that if you want to buy a telescope for use in astronomy it is not wise to pay a large sum for any refractor with an object-glass less than 3 inches in diameter, or a Newtonian reflector with a mirror less than 6 inches across. Smaller telescopes simply do not collect enough light to make them useful. And here we run into the cost problem. In the present age of inflation, prices seem to rise every month, so in January 1988 I took advice from a well-known and reliable telescope-maker. He came up with the following prices:

3-inch refractor, on altazimuth mount: from £380.
3-inch refractor, on equatorial mount: from £500.
6-inch reflector, on equatorial mount: from £430.
$8\frac{3}{4}$-inch reflector, on equatorial mount: from £550.

(As we have seen, an altazimuth is a simple mounting on which a telescope can be moved freely in any direction. An equatorial is more complicated, and means that only one motion is essential: an east-west drive. For astronomical photography, an equatorial is more or less essential.)

Naturally, this is only the beginning. With an equatorial it is possible to add a mechanical drive, usually electrical, which means that the target object will stay fixed in the field of view; but everything has to be very accurate, which means that moving the telescope around is less easy. Obviously, a permanent observatory is the answer, but this is often thwarted by the available horizon, and it always happens that a lofty tree will lie in the most inconvenient direction possible — a case of what astronomers often term Spode's Law (if things *can* be awkward, they *are*). Also, you must have several eyepieces, which adds to the cost. On the credit side, the cost is non-recurring. When you have equipped yourself with a proper telescope, it will last you a lifetime if you take even reasonable care of it.

Then, too, much depends upon your own special interests. If you are concerned mainly with the Moon and planets, then I would be inclined to think seriously about a refractor; for 'deep-sky' objects such as starfields, clusters and nebulae, a reflector is probably preferable. If you want to project sunspots, then go for a refractor every time.

The real trap is that some manufacturers of very small, attractive-looking telescopes make exaggerated claims for them. I have seen a small telescope which was said to bear a power of ×600. It proved to have a $2\frac{1}{2}$-inch objective, so that even if the optics had been of first-class quality the maximum usable power could not possibly have been more than ×125. In 1987 one of the manufacturers was taken to court under the Trades Description Act, and was heavily fined, but many people continue to be taken in. My own guidelines are as follows:

1. Never pay much for a refractor below 3in aperture or a Newtonian reflector below 6in.

2. Avoid any telescope advertisement which gives only the power claimed for the instrument, and says nothing about the aperture of the object-glass or mirror.

3. Always check to see that there is no 'stop' in the tube below the object-glass of a small refractor, which would in effect cut down the aperture still further. (This is a favourite trick, because it tends to conceal faults such as the production of false colour in the image.)

4. Make sure that the mounting is firm. If not, then the telescope will wave about in the slightest breeze, and will be useless.

5. If the advertisement gives a maximum power, check against the aperture of the object-glass or mirror. If the claim is for more than ×50 per inch, avoid the telescope completely.

6. If possible, ask for on-the-spot advice from an expert, or at least someone who has a reasonable knowledge of telescopes. Your local Astronomical Society will generally be able to help here.

I am not suggesting that a very small telescope is of no use at all. Of course, it is better than nothing. What I do say is that in my opinion it is foolish to spend much money on a telescope too small for anything except occasional casual observation. You must be prepared for a substantial outlay. And if this is not possible, then the remedy is to buy a pair of good binoculars.

Binoculars are much more useful astronomically than many people believe. They have most of the advantages of a very small telescope apart from sheer magnification, and of course they can also be used in everyday activities such as bird-watching. They are defined by the aperture of the object-glasses, in millimetres, and the magnification; thus a 7 × 50 pair means a magnification of 7, with each lens 50 millimetres across. In general I would recommend a power of between ×7 and ×10, because with higher magnifications the binoculars have a small field of view and are too heavy to hand-hold properly, so that a mounting becomes necessary. Again I have taken advice on prices as for January 1988, and they are as follows:

7 × 50 binoculars:	from £30.
8 × 30:	from £25.
10 × 50:	from £30.
20 × 80:	from £250.

With the 20 × 80 you will need a mounting, which will cost around £60.

PATRICK MOORE IN 1988, WITH THE OBSERVATORY HOUSING HIS $8\frac{1}{2}$-IN. REFLECTOR

A 'HAND MOUNTING' FOR BINOCULARS Demonstrated for the author by James Savile.

Cheaper binoculars can often be bought second-hand, and the advantage here is that you can test them on the spot before buying; if they will not focus sharply, or if images have any false colour, beware! On the other hand, as I have already said, good-quality second-hand telescopes are very rare. It is worth keeping an eye upon the 'bazaar' column of the British monthly periodical *Astronomy Now*, and the occasional bargain can be found, but again it is wise to take skilled advice if you can.

Of course, you can decide to make a telescope for yourself. Lens-making is a real problem, and probably best left to the professional, but mirrors for reflectors can be made by any skilled amateur. The cost of the 'blanks' for the mirror need not be more than £50 or so, and the mounting is purely a problem of mechanics, but do make sure that you know just what you intend to do before setting out. Alternatively, you can buy the optics ready-made and then mount them.

Once you have equipped yourself with a suitable telescope you will find that you have a source of endless enjoyment, but do think carefully before buying. It is only too easy to make a mistake.

31 | Hipparcos

How far away are the stars? This was a question which baffled astronomers for many centuries — right up to the year 1838, when Friedrich Wilhelm Bessel succeeded in measuring the distance of a rather obscure star in the constellation of the Swan, known by its catalogue number of 61 Cygni. Bessel used the method of parallax; by observing 61 Cygni from opposite sides of the Earth's orbit (thereby obtaining a 'base-line' of 186,000,000 miles) he was able to show that it shifted slightly in relation to the more remote stars. He gave its distance as about 11 light-years, or 60 million million miles.

With stars, there are two motions to consider. One is the annual parallax shift, which is to all intents and purposes inappreciable for stars further away than a thousand light-years or so, and is too slight to be measured accurately beyond a couple of hundred light-years at most. The other is a star's individual or 'proper' motion. Over sufficiently long periods, proper motions of stars will distort the familiar constellation patterns, but even Sirius — which has a very large proper motion by stellar standards, — takes 1,350 years to crawl across the background by a distance equal to the apparent diameter of the full moon. And Sirius, remember, is a mere 8.6 light-years away; there are not many stars closer than that.

We are dealing with very small angles — fractions of a second of arc. One way to demonstrate this is to consider the apparent size of a coin as seen from various distances. Take, for example, the two-shilling piece (or the 10p piece in our modern Mickey Mouse money). Held at arm's-length, the coin will more than cover the Moon; to make it look the same size as the Moon it must be held at 10 feet from the eye. To reduce it to an apparent diameter of a second of arc, it must be taken out to $3\frac{3}{4}$ miles. From Edinburgh it would subtend an angle of 1/100 of a second of arc as seen from London, and from New York a mere 1/1,000 of a second of arc. This is not very much, but it is the sort of quantity with which we have to reckon in stellar parallaxes and proper motions.

In general, Bessel's method can be extended out to measure reliable distances to 80 light-years, and it is reasonable out to perhaps 200 light-years. At 80 light-years, the parallax corresponds to 4/100 of a second of arc. To do better, we must have recourse to space-research methods. This is what is being planned with a new astrometric satellite, Hipparcos. The name is appropriate; it was by comparing the 'modern' star places at the time of Edmond Halley with the positions given two thousand years earlier by the Greek astronomer Hipparchus that Halley was able to show that three bright stars, including Sirius, had shown appreciable proper motion.

Hipparcos stands for *H*igh-*p*recision *P*arallax-*co*llecting *S*atellite. Its main aim is to provide a very accurate map of the positions of 120,000 stars, together with their motions due to parallax and proper motion. The mission is due to last for two and a half years.

Once we know the distance of a star, we can find its intrinsic luminosity. When we also know the colour — which depends on the surface temperature — we can draw up what is called a colour-magnitude diagram in which spectral type is plotted against luminosity. This is actually the same thing as a Hertzsprung–Russell or HR Diagram, because colour and spectral type go together — thus all A-type stars are white, G-type stars are yellow and those of type M are orange-red.

Using Bessel's method, we can draw up a colour-magnitude diagram for stars brighter than apparent magnitude 12 which are within 15 parsecs of us. (One parsec is the distance at which a star would show an annual *par*allax of one *sec*ond of arc; it corresponds to 3.26 light-years. Actually, no star apart from the Sun is within a parsec of us.) This diagram shows only stars of the dwarf or Main Sequence. If we could extend our range to, say, 80 parsecs we would also include stars of the giant branch, which have used up their available hydrogen 'fuel' and have started to burn helium. This would give us a great deal of information about stellar evolution.

The Hipparcos method is analogous to that used by a surveyor in triangulation. The telescope on the satellite has two fields of view, which look at small areas of the sky separated by a fixed angle of 58 degrees. The satellite spins at the rate of 11.5 revolutions per day around an axis which is perpendicular to the two fields of view. During one revolution, a particular star will be seen first in one field and then, twenty minutes later, in the other. The direction of the spin axis is continuously changing, so that the whole sky will be covered many times during the mission.

A finely-ruled grid of 2,688 lines is placed in the common focal plane of the two fields. As a star crosses the grid, taking about 19 seconds to do so — due to the scanning motion — the light signal shows a wave pattern or modulation; the interval between two successive wave-crests is just over 7/1,000 of a second. The location of the star on the grid is given by the phase of the wave. If two stars cross the grid at almost the same time, the angular separation between them is the basic angle of 58 degrees plus the difference in phase between the two waves.

Consider, for instance, the stars in the Pleiades cluster. On each scan, these stars will be measured in relation to stars in many other parts of the sky, and the end result will be a map of relative positions, motions and parallax shifts. It is hoped to achieve an accuracy of 2/1,000 of a second of arc in each of the position co-ordinates, which is about the apparent diameter of our London coin as seen from Moscow. In addition, there is a second experiment on Hipparcos, named the Tycho Project in honour of the great sixteenth-century Danish astronomer who produced the best star catalogue to be compiled before the invention of the telescope. This will provide a virtually complete though less precise survey of the positions, magnitudes and colours of all stars brighter than magnitude 10.

Hipparcos itself is almost ready. The launch was planned for the summer of 1988, but the troubles with the rocket (Ariane) have delayed it until the spring of 1989 at the earliest. Observations should be complete by mid-1992, and the final catalogue is expected to be published in 1995.

It is an exciting and an important project. If all goes well, Hipparcos will provide a first-class source of fundamental data for the benefit of all astronomers in this and the next generation. It will largely supersede the data which have been accumulated through the efforts of astrometrists during the past three hundred years, and it will still be of the utmost value in the far future.

THE NASA HUBBLE SPACE TELESCOPE

32   Cameras and the Sky

Open almost any book on modern astronomy, and you will see magnificent coloured pictures of the Sun, planets and stars. Mars shows its red deserts and its gleaming polar caps; Saturn's gloriously yellow rings stand out; Jupiter's Red Spot is striking; there are coloured stars, vividly hued spiral galaxies, nebulae and much else. The result is that when a newcomer to astronomy first looks through the eye-end of a telescope, even a large one, he is apt to be disappointed. For example, the Andromeda Spiral, a vast star-system larger than our own, and so remote that its light takes over two million years to reach us, looks like nothing more than a dim smudge of light without any well-defined shape, and there is no colour at all. It is difficult to realize that the superb photographs so often published, and shown on television, are genuine.

The troubles are twofold. First, large instruments and highly sophisticated equipment are needed to take good pictures of objects such as spiral galaxies, mainly because by everyday standards they are so faint; the Andromeda Spiral is barely visible with the naked eye even when you know where to look for it. Secondly, the colours — real though they are — are very fugitive. Remember, the intensity of light has to reach a certain value before the colours become visible. If you doubt this, go out on a night when the Moon is full and try to tell the difference in hue between a red car and a green one; you will fail, because the moonlight is not sufficiently brilliant, even though it may seem so. (In fact, it would take around 400,000 full moons to match the light of the Sun.)

Quite apart from this, professional astronomical pictures often have false colour added to them to make analysis easier. The classic examples of such a procedure were the pictures of Halley's Comet transmitted during the period when the Giotto space-craft made its rendezvous, in March 1986. Most viewers expected to see something which looked like a conventional comet; instead, they were treated to what appeared to be gaudy patterns. This was scientifically sensible, as the different colours represented levels of different light intensity, but it is not surprising that many people were confused. From my improvised BBC studio in Darmstadt, headquarters of the European Space Agency, I had to try to explain what was happening!

It is quite possible for really skilful and well-equipped amateurs, who own large telescopes, to take photographs which are of professional standard. On the other hand, there is also scope for the complete novice, and this is something which is not always appreciated. All you really need is a camera which can be set to give time-exposures. The results will be aesthetically pleasing, and may even be of distinct scientific value.

Because all celestial objects apart from the Sun and the Moon are comparatively dim, a time-exposure is essential. The camera must be fixed upon a really firm tripod, and to avoid jerking it you need that simple appliance known as a cable release. The immediate problem to be faced is that the Earth is spinning on its axis, so that the entire sky seems to be moving round from east to west, carrying the celestial bodies with it. If, therefore, you point your camera at a starlit sky and give an exposure of several minutes, you will find that the stars are drawn out into trails. The only way to avoid this is to drive the camera round at a rate just sufficient to compensate for the Earth's rotation, and there are immediate complications for the would-be photographer who has no extra equipment.

However, star-trails themselves can be attractive. A wide-angle lens is to be recommended, and, obviously, a fairly fast film; if you use a colour film with ASA (or, in modern parlance, ISO) 400 you should obtain good results. It is particularly interesting to aim the camera at the Pole Star, which lies within a degree of the north pole of the sky and seems to remain almost motionless with everything else revolving round it. With an exposure of a quarter of an hour or so, you will see that the stars are drawn out into graceful curved arcs, and even Polaris itself shows a short curve. Limitations are imposed by 'fogging' due to sky brightness, and artificial lights are the sworn enemy of the astronomical photographer; you must also beware of possible dewing-up of the optics. Try a series of exposures ranging from, say, a minute or so up to half an hour. With exposures of less than about 30 seconds there is not much trailing of the star images.

It is easy to make a mistake. Flash a torch anywhere near the camera while the exposure is being made, and you will end up with a blank picture. Aircraft can also be a menace, and can easily spoil what would otherwise be a good photograph. However, after a few attempts you will be unlucky not to produce something which will please you.

Now and then phenomena occur which simply ask to be photographed. In 1987 the brilliant planets Mars and Jupiter were in conjunction in the evening sky, and I took a series of pictures, not because I thought they would be of any use but simply for the fun of it. One of them was taken from inside my observatory, through the slit in the dome, just before the two planets dived behind an inconvenient tree (not in my garden; if it had been, it would have met with a sad fate long ago). The exposure was only one minute, so there was little trail.

With longer exposures, trail becomes obtrusive, but the results are still pleasing. In 1986 I found myself in the Southern Hemisphere, looking at Halley's Comet. The main southern constellation is Crux, the Cross, which is actually more like a kite than an X. I photographed it with a fast film (1,000 ASA) and a two-minute exposure. The pattern came out well; note that of the four main stars three are pure white, while the fourth, Gamma Crucis, is obviously orange-red. This shows that Gamma Crucis has a relatively cool surface. It is an older star, which has used up its main fuel and is approaching senility. Halley's Comet itself was never bright, but it was not difficult to record with an ordinary camera on a fixed tripod.

You may also be lucky enough to snare a meteor as it flashes across the field of view. Meteors, tiny particles of sand-grain size burning out in the Earth's upper air, can be bright; you cannot predict them, but there are various well-defined annual showers, and if you aim the camera towards the radiant of the shower (the point in the sky from which the meteors appear to come), you should be reasonably optimistic. August is the best month for this, because between the end of July and about 17 August the Earth passes through the meteor shoal known as the Perseids, and bright shooting-stars are common.

Still with an ordinary camera, what about aurorae, or polar lights? These lovely displays are best seen in high latitudes, so that they are very frequent from Scotland or North Norway but much less so from southern England, while from countries near the equator they are hardly ever seen. They are due to electrified particles sent out by the Sun, which cascade into the Earth's upper atmosphere and produce glows. The colours are often vivid, and aurorae change rapidly. With a fast film, a short exposure will bring them out well.

Aurorae are commonest when the Sun is active. There is an eleven-year cycle of activity, and we are now starting to build up to the next maximum, so that there should be plenty of aurorae between now and around 1992. From England or South Europe, you may have to wait; but if you go further north, you should have no problem in seeing — and capturing — an aurora. Again, try a series of exposures, ranging from a few seconds up to several minutes depending upon the brightness of the display.

With a 35 mm camera and 400 ASA film, the Moon is a promising subject. Again there is trail after a few seconds, and with an ordinary lens, suitable for daylight use, the Moon will seem very small; but if you have a telephoto lens you should be able to record the main markings, such as the dark, waterless 'seas' and some of the mountains and craters. It is wise to bracket the exposures, beginning with some which are probably too short and working through to those which are probably too long; some of the pictures should be about right.

New problems arise when you are trying to photograph several objects in the same area which are different in brightness. For example, if you try for the crescent Moon and a planet together you must either expose for the planet, thereby over-exposing the Moon, or else expose for the Moon, when the planet may not show up at all. It is all a question of trial and error.

When taking astronomical pictures with an ordinary camera, do not forget that commercial firms are unused to them, and may reject a picture which looks to them like a black area with a few blobs on it. It is wise to send a letter of warning — and also to begin the film by taking an ordinary everyday picture, so that when the film is cut up your astronomical pictures will not be sliced in half.

Colour photography is now the norm, but black-and-white pictures are not to be despised, and if you process them at home you can work wonders by varying the developing methods.

To take longer exposures without trailing, the camera must be guided. Simple devices can be made to allow for this, or the camera can be mounted upon a driven telescope, so that the telescope merely acts as a guide. I well remember taking my first picture of Halley's Comet, on 14 October 1985, when it was still very faint. I fitted my camera on to the telescope, identified the star-field and took a ten-minute exposure. When I developed it, I found that I had recorded the comet as a faint speck. There was also a streak across the picture, due to a Russian artificial satellite which discourteously crawled across the field while the exposure was being made. Had it been slightly displaced, it would have passed right in front of the comet and blotted it out.

Another good subject for photography by this method is the Orion Nebula. You can see it with the naked eye as a misty patch; it is a stellar nursery, over 1,000 light-years away, made up of dust and gas lit up by the hot stars mixed in with it. With a 1,000 ASA film I took a wide-angle picture of it in only three minutes, and the colour showed up well — much better than it ever does visually.

I do not claim to be a serious astronomical photographer, and this is no place to describe techniques of taking pictures through a telescope, or even recording spectacular phenomena such as solar eclipses. But it can be done, and if you want to try your hand there are many books to help you. There is nothing really difficult about it — and it is always satisfying to produce one's own pictures, even if you will be unable to bring out the forms and vivid colours of nebulae and galaxies far away across the universe. I wish you all success.

33   The Hot Oceans of Venus

It is not surprising that the ancients named Venus in honour of the Goddess of Beauty. When at its best, as it was during the winter of 1987–8, the planet is indeed a glorious sight. It is far brighter than any other object in the sky apart from the Sun and the Moon, and it can even cast a shadow. Keen-sighted people can glimpse it in broad daylight, and it is easy enough to see immediately after the Sun has dipped below the horizon.

Telescopically, it must be admitted that Venus is something of a disappointment, because all that can be made out are vague shadings which look — and are — mere cloud phenomena. Nobody has ever had a direct view of the surface, because the clouds never clear away; if you were standing on Venus, you would have no chance of seeing the Sun. This meant that before the Space Age, our knowledge was relatively meagre. We knew that Venus is very slightly smaller and less massive than the Earth; that its 'year' is 224.7 Earth-days; that it shows phases, like those of the Moon, and that its upper atmosphere contains large amounts of the heavy, unbreathable gas carbon dioxide. That was really about all. We did not know how long Venus takes to spin on its axis, though the best estimates gave a period of around one month.

Carbon dioxide acts in the manner of a greenhouse, and tends to blanket in heat. Since Venus is closer to the Sun than we are (67,000,000 miles, as against 93,000,000 miles for the Earth), it was thought that the surface must be hot; but *how* hot? According to F. L. Whipple and D. H. Menzel, Venus was probably a marine world, with almost no dry land. Since the atmospheric carbon dioxide would have penetrated the water, the result would therefore be seas of soda-water. Sir Fred Hoyle preferred seas of oil; others believed Venus to be bone-dry, without a trace of moisture anywhere.

There were minor mysteries, too. On occasions, when Venus is a crescent, the 'night' side can be seen shining dimly. The same effect is visible with the crescent Moon, but is easy to explain; it is due to light reflected on to the Moon from the Earth. Venus, however, has no moon, and the 'Ashen Light' is much less easy to account for. A last-century German astronomer who rejoiced in the name of Franz von Paula Gruithuisen suggested that it might be due to illuminations on the planet's surface, set up by the locals to celebrate the election of a new Government, but on the whole it seems rather more likely that it is

ARTIST'S IMPRESSION (BY PAUL DOHERTY) OF THE LONG-VANISHED HOT OCEANS OF
VENUS

caused by electrical effects in Venus' upper atmosphere. In any case, the chances of life on Venus appeared to be nil. Whether life had developed there in the remote past was another matter.

Then, in 1962, the first successful interplanetary probe — America's Mariner 2 — flew past Venus at around 20,000 miles, and sent back a great deal of information, not all of which was entirely welcome. First, the rotation period was long; we now know it to be 243 Earth-days, longer than Venus' 'year', and the direction of spin is east to west, not west to east. The reasons for this remain unknown. Second, and more importantly, the surface proved to be fiercely hot, at something like 900 degrees Fahrenheit. And thirdly, the clouds were rich in sulphuric acid. Together with a carbon-dioxide atmosphere with a ground pressure 90 times that of the Earth's air at sea-level, all this made Venus unsuitable as a holiday resort.

Later the Russians soft-landed probes, and obtained pictures direct from the surface, showing rocks which glowed orange in the intense heat. Also, both the Russians and the Americans were able to map the surface by radar, partly from Earth but mainly from space-craft put into closed orbits round Venus. There are two major highlands (Ishtar and Aphrodite), craters, valleys, a huge rolling plain, and what appear to be active volcanoes. In the area now called Beta Regio there are two massive shield volcanoes, Rhea Mons and Theia Mons, both of which

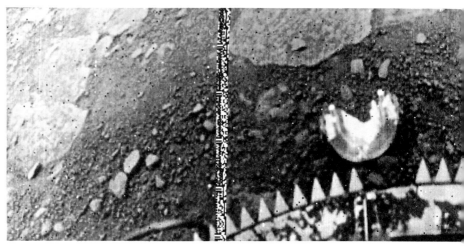

PART OF THE SURFACE OF VENUS Taken by the descent module of the Soviet Venera 13 space-craft. This module soft-landed on 1 March 1982.

are far more impressive than our shield volcanoes such as Mauna Kea and Mauna Loa in Hawaii.

On Earth a volcano is produced by a 'hot spot' below the crust. But according to the modern science of plate tectonics, the crust is drifting slowly over the mantle; and when a volcano moves away from the hot spot it becomes extinct. This is what has happened to Mauna Kea, which has not erupted for many centuries, and will presumably never do so again (at least, I hope not, because one of the world's major observatories has been erected on its summit). Mauna Loa, its neighbour, is highly active, but in time it too will drift away and cease to erupt.

On Venus the crust appears to be thicker than that of the Earth, and there is no steady drift, so that once a volcano has been formed it will continue to erupt for a very long time. This is why Rhea Mons and Theia Mons are certainly much more massive than anything we find on Earth, even in Hawaii.

So what is Venus really like? With an atmosphere made up chiefly of carbon dioxide, and with its tremendous temperature and surface pressure, it is as hostile as could be imagined. There are constant eruptions, and probably continuous thunder and lightning; all in all, Venus is remarkably like the conventional picture of hell. Yet it is so like the Earth in size and mass that, logically, it ought to be in the same state.

As we have noted, the difference must surely be due to the fact that Venus is over twenty million miles closer to the Sun. Up to now it has been assumed that the carbon-dioxide atmosphere produced what may

be called a runaway greenhouse effect. In the early history of the Solar System — more than 4,500 million years ago — the Sun was less hot than it is now, and Venus and the Earth began to evolve along similar lines. But as the Sun became more energetic, the oceans of Venus boiled away, the carbonates were driven out of the rocks, and any life which had appeared there was ruthlessly snuffed out.

Yet what has happened to the water? There is virtually none left now, and this has led to a new 'wet greenhouse' theory due to three American astronomers: James Kasting, Tom Ackerman and James Pollack.

We begin, as before, with the assumption that at an early stage Venus and the Earth were very alike. As the Sun heated up, Venus became so hot that the water evaporated, and this process went on until 50 per cent of the planet's atmosphere was water vapour. Then, however, evaporation stopped, because the water-vapour pressure in the atmosphere had become too great. The remaining water stayed liquid, but hydrogen continued to escape from the top of the planet's atmosphere — to be slowly replenished by new evaporation from the oceans.

By remaining liquid for millions of years, the oceans were able to move the original huge mass of carbon dioxide gas out of the atmosphere, converting most of it into carbonate rocks in the planet's crust ('weathering', if you like). This reduced the density of the atmosphere until it was no thicker than ours is today. By the end of this process, nearly all Venus' water had been lost, and the oceans were no more. Only later did the atmospheric pressure build up again.

VENUS 0463   80 090   1542   F 569

VIEW OF VENUS FROM THE PIONEER ORBITER

Speculative? Certainly; but at least it would explain the present lack of water. It has been converted into hydrogen and oxygen; the hydrogen was then lost in space, while the oxygen combined with crustal materials to make up oxygen-rich minerals such as haematite and magnetite.

If Venus had oceans for several hundred million years, life might have started there, and persisted until the oceans were almost at boiling-point. It conjures up a strange picture, with the hot, bubbling seas covering much of the planet and the living organisms in them trying to resist the steadily-increasing temperature. Eventually they lost the battle, and from being a life-bearing world Venus changed into a sterile globe.

If this picture is right, then we cannot entirely rule out the possibility of finding traces of past life in Venus' rocks — once we can analyze them. Obviously, the chances of a manned mission there can be discounted, but there is no reason why we should not manage a sample-and-return probe within the next decade or two.

My own view — a personal one only — is that we are unlikely to find any 'fossils' in the rocks of Venus, but I rather hope that I am wrong. It would be fascinating to look back several thousands of millions of years to a world where the last living creatures were battling to survive in their boiling, ultimately deadly ocean.

34 The Yerkes Observatory

We live in the age of great telescopes. New giant instruments are being planned, and some of these will be far superior in light-grasp to the famous Palomar 200-inch reflector, which for decades following its completion in 1948 was incomparably the most powerful telescope in the world. But what about that other type of telescope, the refractor? The first astronomical telescopes, constructed around 1609, were of this type; it was not until some sixty years later that Isaac Newton built the first reflector.

Just as a reflector uses a mirror to collect its light, so a refractor uses a lens or object-glass, which today is usually made up of several components. Inch for inch, a refractor has the advantage; thus a 3-inch refractor is powerful enough to be useful, whereas a 3-inch Newtonian reflector is, to be frank, fairly useless. Also, refractors are not temperamental, and they need little maintenance, On the other hand, they are more expensive than reflectors of comparable light-grasp, and there is a limit to their size. Remember, a mirror for a reflector can be supported on its back; but with the objective of a refractor, the light has to pass straight through, and the lens has to be supported all round its edge. If the lens is too large, it will also become too heavy, and will start to distort under its own weight, thereby ruining its performance.

This is why the modern emphasis is upon reflectors (plus the fact that the light-correction is better; even the best refractor tends to give a certain amount of 'false colour', which is why Newton, for one, had no faith in them). A century ago the outlook was different. In 1888 a 36-

inch refractor was completed at the Lick Observatory in the United States, and nine years later came the completion of what is still the world's largest refractor, the 40-inch at the Yerkes Observatory at Williams Bay, near Chicago.

The telescope was the brainchild of a remarkable man: George Ellery Hale. Basically he was a student of the Sun, and during his career he made some discoveries of vital importance, notably the magnetic fields of sunspots; he was also the inventor of the spectro-heliograph. He was also interested in what we now call 'deep space', and he knew that for this sort of work one needs a telescope with tremendous light-grasp. Why not build a refractor bigger than the Lick 36-inch?

Money was needed — a great deal of it — and though Hale and his family were well off, they were not millionaires. So Hale cast around for a friendly millionaire, and he found one in Charles Tyson Yerkes, a Chicago street-car magnate. Hale's sales-talk was excellent, and in the autumn of 1892 Yerkes agreed to provide half a million dollars for the construction of what would be the world's most powerful telescope. Yerkes insisted that it should be sited within a hundred miles of Chicago, which is why Williams Bay, on the shores of Lake Geneva, was chosen.

The lenses were finished in the summer of 1895. In May 1897 they were shipped to Lake Geneva, and mounted in the tube, which is over 60 feet long. Tests were made, but there was a near-disaster almost at once. The observatory had been equipped with an elevator-type floor to make it easy for the observer to reach the eyepiece of the telescope; the floor weighs $37\frac{1}{2}$ tons, and during the testing the cables holding it snapped. The floor crashed to the ground, and it was said that the inside of the dome looked 'as if a cyclone had slipped through the slit and gone on the rampage'. Mercifully, nobody was hurt, and the telescope was finally dedicated on 21 October 1897.

The 40-inch soon proved to be every whit as good as had been hoped. It has never been surpassed in size; admittedly a 49-inch was once built and given some preliminary tests in France, but it was never used seriously and was never set up in an observatory (nobody now seems to be sure what happened to the lens). So the Yerkes refractor remains in a class of its own, and now, after almost a century, it is still in use on every clear night. Of course, it is not the only telescope at Yerkes; there is also a 24-inch reflector, together with various smaller instruments.

I cannot resist recounting two somewhat irreverent stories about the early days at Yerkes. The dedication did not pass off without incident. The Observatory building had been cleverly designed, and at the main entrance were two ornamental stone columns with elaborate patterns. Just before the dedication, a member of the Observatory Board noticed that one of the designs indicated a man about to be stung on the nose by a bee. Most undignified! Moreover, could the man represent Charles Tyson Yerkes, being 'stung' for the money? This would not do at all; so a stonemason was called in, and solemnly chiselled away ninety-five bees. You can still see the bare patches where the bees used to be … Then there was Edwin Frost, who succeeded Hale as Director. Frost worked out a new way of telling the temperature without consulting a thermometer. There are many crickets around the Observatory; all you do is to count the number of chirps made by a cricket in thirteen seconds, add 40, and — hey presto! — you have the temperature in degrees Fahrenheit. I have always longed to check this personally, but whenever I have been to Yerkes it has always coincided with the close season for crickets!

The 40-inch itself has a total weight of twenty tons, but is so perfectly balanced that it can be moved by hand. Today it is used largely for parallax measurements of stars, for which it is ideally suited. Also, the work goes back a long way, and it is of great value to compare photographs taken decades ago with those of the present time taken with the same instrument. For example, Dr Kyle Cudworth has been studying globular clusters and comparing the photographs of the 1900 period with those of the 1980s; the tiny shifts of individual stars provide a tremendous amount of information. It was with the 40-inch, too, that Dr W. W. Morgan obtained results enabling him to show that our Galaxy is indeed a spiral system, like a vast Catherine-wheel. Morgan has been at Yerkes for sixty years now, and was Director for part of this period; other famous Directors have been Otto Struve and Gerard P. Kuiper. Today, new equipment is being developed and tested there for use in space vehicles.

Most major observatories are very much off the beaten track. Mauna Kea is 14,000 feet high; even Palomar is not too easy to reach, but Yerkes is an easy drive from Chicago, and it actually adjoins a golf-course. Rather surprisingly, the seeing conditions there are still quite good, despite the increase in light-pollution. Moreover, the 40-inch needs little maintenance; unlike reflectors, it is the reverse of temperamental.

I last went to Yerkes in 1986. Drive up to it, and it may strike you as being a little old-fashioned; there is no other observatory of similar status where the main telescope is a refractor. But this is highly misleading. By the very nature of its equipment, Yerkes can undertake observational programmes which other establishments would find difficult. And the telescope itself is immensely impressive; unlike the

OVERALL VIEW OF YERKES OBSERVATORY The dome of the 40-in. refractor is to the left. Taken by the author in 1987.

familiar skeleton-tube reflectors, it really *looks* like a telescope. You do not even have to go outdoors to enter the dome; there is direct access from the main building.

As a lunar and planetary observer, I was naturally very interested in the Solar System observations which have been made from Yerkes over the years. One leader in this field was the late John E. Mellish, and it was he who was involved in a truly remarkable episode which also involved another eminent observer, Edward Emerson Barnard. Both these two were renowned for their keen sight, and both looked carefully at the planet Mars.

Around this period — the time of the First World War — the 'canal' controversy was still very much in the public eye, and there was definite backing for Percival Lowell's contention that the canals were artificial. Lowell had recorded them with the 24-inch refractor at Flagstaff. Neither Barnard nor Mellish could see them — and Barnard had earlier been equally unsuccessful with the Lick 36-inch. But then, in November 1915, Mellish looked at Mars, using magnifications of 750 to 1100, and saw — not canals, but craters: hundreds of tiny, circular objects standing out in relief. One of them was estimated to have a diameter of 250 miles. Yet at that stage nobody had had any idea that there might be craters on Mars.

Why had Lowell missed them? Mellish stopped down the aperture of the 40-inch to a mere 24 inches. The craters disappeared. Evidently,

THE 40-INCH REFRACTOR AT YERKES Photographed by the author.

165

then, they were beyond the receiving power of Lowell's telescope.

Mellish consulted Barnard, and asked: had Mars really showed 'many craters and cracks and mountains'? Barnard laughed, and produced drawings which he had made in 1892 and 1893 with the Lick telescope. There were plenty of craters, but Barnard had kept quiet because he felt that 'no one would believe him, and others would only make fun of it'. Those drawings cannot now be found, while Mellish's sketches were lost in a disastrous fire in 1964 . . .

What can we make of this? Both Barnard and Mellish were brilliant observers of unquestioned honesty. There is no doubt in my mind that they did see craters; anything else is unthinkable. I was very anxious to look for myself, and luckily I have had the chance to do so.

In the 1950s and most of the 1960s I was engaged in official mapping of the Moon, and I spent some time at Flagstaff using the 24-inch. I was able to look at Mars quite often, under good conditions, but I could see neither craters nor canals. Neither could I do so with other great refractors, such as the 33-inch at Meudon, near Paris, and the 27-inch at Johannesburg, both of which I know well and have used extensively. But it could have been that an even larger aperture was needed, as Mellish had supposed. The Lick 36-inch is now more or less out of commission, but the 40-inch is as good as ever.

In October 1986 Mars was on view; the apparent diameter was 12 seconds of arc — greater than it had been in November 1915, when Mellish made his observations. I was at Yerkes, presenting a *Sky at Night* programme. Before the routine observations with the 40-inch were started on the night of October 11, I asked Dr Al Harper, the Director, whether I could monopolize it briefly to look at Mars. He kindly agreed. Using a magnification of 1,000, Mars came into the field of view — the first time I had seen it through the 40-inch.

Did I expect to see craters? Frankly, I am not sure. In the event, the main surface features stood out magnificently; so did the polar cap; there was no obvious dust-storm activity on the planet. Try as I might, I could not see craters. Had they been even reasonably obvious, I am confident that I would have done. So my education, as far as it goes from a single observation, is that to see the craters of Mars you need not only the power of the 40-inch, but also the eyes of a Barnard or a Mellish.

I left Yerkes with regret. It is a friendly place — our television team was given every help and courtesy by the Director and his staff — and it is unique. It has a long and honourable history extending now over almost a century, and I have not the slighest doubt that in a hundred years from now it will remain where it is today: in the forefront of astronomical research.

35 Totality from Talikud Island

Have you ever heard of Talikud Island? I admit that I had not, before the early part of 1988. Then I found that it is quite small (a few miles long and wide) and is in Davao Bay, in the Philippines, within 6 degrees of the equator. Davao is the main city of Mindanao, the southernmost of the larger Philippines — the only other major town there is General Santos City — and it is decidedly unstable politically, which is why many people steer clear of it.

What brought it to my attention was that Davao lay right in the centre of the track of the total solar eclipse of March 17–18 (the eclipse extended over midnight GMT). The Explorers Travel Club was mounting an expedition there. I was asked to take part, and this seemed to be an excellent idea, so on March 12 I rendezvoused with the other members of the party at Gatwick and boarded the plane for Manila.

We travelled by Philippine Airways. Once in the air there are no real problems, but their idea of timekeeping is not really precise, and a delay of three or four hours is regarded as being quite punctual. However, we reached Manila eventually; we went to the local Planetarium to hear a lecture delivered by a speaker whose voice reminded me strongly of a duck in the distance, and then we flew on to Davao.

The Filipinos do not believe in doing things by halves. Instead of one set of terrorists, they have three. Of these, the NPA (National People's Army) is fairly conventional, and is modelled on the lines of the IRA or the ANC. The MNLI is trying to liberate somebody from something, and there are also supporters of ex-President Ferdinand Marcos, who was overthrown a few years ago by the current President, Mrs Corazon Aquino. All these groups tend to blow things up. They also blow each other up, but on this occasion they were kind enough to put out a combined Press statement to the effect that they did not propose to throw any bombs at the visiting astronomers. We felt reassured, and in the event we heard gunfire only once!

When we reached Davao the temperature was of the order of a hundred in the shade. Davao itself is a fair-sized town — larger than General Santos City — and we had chosen it because we expected conditions to be at their best there. The plan was to carry out the actual observations from Talikud Island, in the Bay, because any cloud was

167

liable to hug the main coastline. Talikud is a typical Pacific island, with palm-trees and glorious beaches; a few people live there, but not many. You can see it easily from Davao, and to reach it you sail across the Bay in rather flimsy outrigger boats.

The members of our party had various types of equipment, ranging from the elaborate (Bob Turner, Douglas Arnold and others) to the rudimentary (me, with my 400mm telephoto lens on a somewhat elderly Pentax camera). There were more than fifty of us all told. On 16 March we made a reconnaissance trip, and selected our sites; we decided upon the beach, so that the eclipsed Sun would (we hoped) show up over the palm-trees, and the Moon's shadow would come towards us from across the sea.

The only problem was the weather. Clouds gathered, and on the morning of 'eclipse day' it was overcast. Should we go to Talikud Island, as planned, or stay in Davao, or try for General Santos City? In the end most of us opted for Talikud, and at 5 a.m. Filipino time we set off. When we reached the island it was still cloudy, and we arranged our equipment more in hope than in expectation.

First Contact — and still not a break in the clouds. The gloom increased. Totality was due at 9.07 Filipino time, but by half-past eight the situation was looking decidedly grim. It was worse by 8.45, when a gentle rain started to fall, and Bob Turner hastily covered his telescope with a large, multicoloured umbrella which he had been wise enough to bring along.

I think that most of us had abandoned all hope by 8.55. Little more than ten minutes to go; the light was fading, but still no break. I wondered what was the best course. Should I take my camera off its tripod, put in my wide-angled lens, and simply try for the Moon's shadow? Luckily, I decided against it; hope springs eternal. Then the rain stopped, and to my delight I saw the thin crescent of the Sun. Hastily I pointed my camera in what I hoped was the right direction, and took an exposure at 1/250 of a second (I was using professional 100 ISO film). The cloud was thinning, but another dark mass was approaching, and I offered up a prayer that it would not arrive before totality.

Abruptly the light faded. Glancing over my shoulder, I saw the Moon's shadow racing towards us across the sea; it was more dramatic than the shadow I had seen at any of my five previous totalities. There was the flash of the Diamond Ring, and, miraculously, there was the eclipsed Sun.

It was amazing. A certain amount of light remained, and this hid the outer corona, but the inner corona was bright, and there were two glorious red prominences. I suppose that the overall effect was

TOTALITY As seen from Talikud Island. The clouds cleared shortly before totality.

enhanced by the idyllic scene below, though the light was dim; it was as though the thin cloud enhanced the prominences at the expense of the corona. I doubt whether either Venus or Jupiter could have been seen, but I had no time to look. Everything was still: Nature had suddenly called a halt.

We had three minutes, and I took a series of exposures. Since I had had no preparation time to speak of, all I could do was to bracket from around 1/10 second up to around 4 seconds, and hope that one picture would be all right. There was no alternative, because it was impossible to gauge the effects of the residual cloud. Mercifully, the dark mass kept away, menacing though it looked. The tension remained until Third Contact, with the flash of the Diamond Ring just before the light flooded back over Talikud Island. We had been incredibly lucky.

I went on photographing the post-totality partial phase until the clouds came back again, this time permanently, at around 9.25. Then we compared notes. Some of us had been unfortunate, because of the impossibility of 'lining-up' until the very last moment, and there had been two cases of jammed cameras. All in all, the results were patchy; I hoped that I would have one good picture (as, in the event, I did).

Packing up was a relatively joyful process. After all, when we had landed on the island a few hours before none of us really thought that we would see anything at all. We said farewell to Talikud with considerable affection. Next day we returned to Manila, went to the Observatory (where we were made very welcome by the Director, Dr

Badillo, and his colleague Father Haydon) and even went on a sightseeing trip to Taal, one of the world's really active volcanoes, which you have to reach by boat.

Was this my most spectacular eclipse? In some ways I think it was. Conditions for photography were far from ideal, but the sight of that gloriously-eclipsed Sun, with its prominences, shining down from above the palm-trees is something that I will never forget.

Subsequently I found that I had had one more piece of personal luck. Mrs Aquino, President of the Philippines, had travelled to General Santos City to see the eclipse. I had been expected there (I am not sure why) and apparently I was to have an official invitation to join the President's party. Sadly, clouds came up at the critical moment, and Mrs Aquino saw nothing at all. Therefore, I cannot complain!

36 Mysterious Sun

The 1988 solar eclipse was spectacular by any standards. As I have related, from Talikud Island in the Philippines I was able to see the Sun's corona and prominences with the naked eye. But how much do we really know about the Sun, and how much do we yet have to find out?

We know, of course, that the Sun is an ordinary star, around 865,000 miles in diameter (big enough to swallow up more than a million Earths) and that it is 93,000,000 miles away. It shines not because it is burning, but because of nuclear reactions going on near its core, where the temperature is at least 14,000,000 degrees Centigrade and the pressures are colossal. Broadly speaking, hydrogen is being converted into helium, with release of energy and a mass-loss at the rate of 4,000,000 tons per second. The Sun is around 5,000 million years old, which by stellar standards is middle-aged.

On its surface we can often see sunspots, which are patches where the temperature is around 2,000 degrees cooler than normal. Individual spots range from tiny pores through to major groups over 50,000 miles long and covering areas of hundreds of millions of square miles. A major spot has a dark central region or umbra, surrounded by a lighter penumbra where the temperature is of the order of 5,500 degrees. Groups may be highly complex, with many umbrae immersed in an irregular penumbral mass.

Spots are centres of strong localized magnetic fields, with strengths

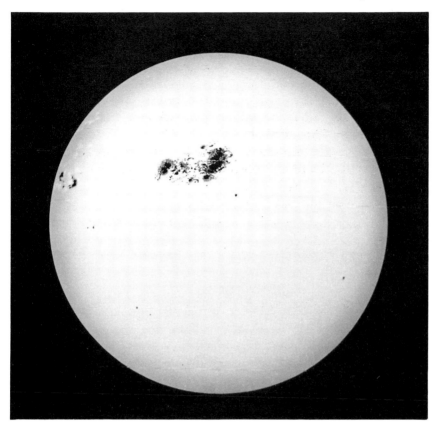

THE GREAT SUNSPOT GROUP OF 1947 — THE LARGEST EVER SEEN — DRAWN BY
PAUL DOHERTY

up to 10,000 times that of the Earth's field. These magnetic fields are
revealed by the spectroscope. The Sun's spectrum consists of a
rainbow or continuous background crossed by large numbers of dark
lines, each of which is characteristic of one particular element or group
of elements. If a magnetic field is present, each line is split into several
components; this is the so-called Zeeman Effect, and is very evident
with the dark Fraunhofer lines in the spectrum of the Sun.

Magnetic forces seem to be responsible for controlling almost all
aspects of solar activity. Each sunspot pair or group is a bipolar
magnetic region. If one spot of the pair behaves as a north magnetic
pole, the other will behave as a south pole, just as though there were a
bar magnet embedded just below the spots. It seems that we are
dealing with a region where a loop of sub-surface magnetic lines
penetrates the surface. Where the lines enter you have a spot of one
polarity; where they emerge you have a spot of opposite polarity.

Convection seems to be the driving force in concentrating the Sun's

171

magnetism into localized areas. We are all familiar with convection, where hot matter rises to the surface of a gas or liquid, spreads out, and then cools and sinks before being re-heated once more. We can see direct evidence of convection in the short-lived granules of hot gas which give the photosphere its 'rice-grain' aspect. The granules are simply convection cells, each about 1,600 miles across; the bright regions represent hot gas rising and spreading out, while the darker, cooler regions represent lower-temperature gas which has started to sink. Larger cells called supergranules have also been identified: these are around 20,000 miles in diameter and 6,000 miles deep. Still larger giant cells or hypergranules also exist, dredging up material from perhaps 130,000 miles below the visible surface.

The nuclear reactions which power the Sun are taking place in the core, which extends out to approximately one-quarter of the radius of the globe. Beyond this is the radiative zone, through which photons of energy battle their way, suffering thousands of millions of collisions as they struggle towards the surface. The outermost 3 per cent of the solar radius makes up the convective zone, where energy is transported to the photosphere solely by convection. The powerful convection currents push the lines of magnetic force together, and this concentrates them into tubes, or twist structures known as ropes. The density of the gas inside these tubes becomes less than that of the surrounding material, and therefore the tubes become buoyant and float towards the surface as soon as the magnetic fields have become strong enough. Where they break through a spot or bipolar magnetic region is formed. The spots are relatively cool because their strong magnetic fields restrict the circulation of gas in the convective cells nearby, and so reduce the amount of heat energy flowing out through the region occupied by the sunspot.

Next, let us consider the solar cycle. Every eleven years or so the Sun is at its most active, with many spots and groups, while at minimum there may be many consecutive spotless days, as was the case in 1986 and 1987. At the start of a new cycle, the first spots appear at around latitude 30 degrees, while spots of the old cycle may still be seen nearer the Sun's equator. As the cycle progresses, the zones where spots occur migrate towards the equator, reaching latitude 15 degrees at maximum activity. Thereafter, as the numbers diminish, the spots appear progressively closer to the equator until the cycle fades out and the next one begins. It is also notable that the magnetic field reverses at the end of each eleven years. Thus if the leading spots in the Sun's northern hemisphere had north polarity in one cycle, the leading spots in the same hemisphere will have south polarity in the following cycle. It takes 22 years for conditions to revert to their original state, so that, strictly speaking, the solar cycle amounts to 22 years rather than eleven.

We have to admit that as yet there is no theory of the solar cycle which is completely satisfactory. Unlike the Earth's magnetic field, which is generated by currents circulating deep in the core, the solar field may be described as no more than skin-deep; it seems to be due to a dynamo process involving circulating currents which are driven by convection in the convection zone. Horace Babcock's theory links this shallow field with the Sun's differential rotation. The rotation period at the equator is 25 days, increasing to 27 days at latitude 30 degrees and 34 days at the solar poles. Thus if we imagine a row of spots aligned along the north-south line, then after one rotation the spots nearer the equator will have 'pulled ahead' of the others. The Sun's surface is in a continual state of twisting.

If we now assume lines of magnetic force running due north-south below the Sun's surface, the differential rotation will distort the lines. After several rotations, the lines will have become wrapped round the Sun and will be concentrated together. The more they are concentrated, the stronger becomes the magnetic field, and eventually the lines form tubes which penetrate the surface, first at around 30 degrees latitude and subsequently closer to the equator. If we follow the line of force from one hemisphere to the other, we can now see why this model leads to spot-pairs with opposite polarities in opposite hemispheres. As each group decays, the magnetic polarity associated with the following member of the pair tends to drift preferentially toward the Sun's pole, thereby reversing the overall magnetic field and starting a new cycle with the polarity pattern reversed.

There is, however, a more recent theory, due to Peter Wilson and Herschel Snodgrass. There is evidence that two cycles are operating simultaneously; the Sun shows what are called torsional oscillations — i.e., each hemisphere tends to be divided into four regions, where the rotation speed is alternately slightly faster and slightly slower than the mean for those latitudes. These zones migrate from pole to equator in 22 years, and spots tend to appear near the boundaries between a fast and a slow zone when that zone first reaches about 30 degrees from the equator. Short-lived, small-scale bipolar regions, which do not produce spots, form in high latitudes with polarity patterns opposite to the visible spots, and they too migrate equatorward, reaching latitude 30 degrees some eleven years after the beginning of the preceding sunspot cycle. The motion of bright patches in the solar corona — patches which are reflections of the underlying magnetic fields — from polar regions towards lower latitudes indicates a similar trend. This behaviour may be caused by giant convective rolls which form at the poles and gradually move towards the equator, breaking up as they go and taking 22 years to reach the equator. Four such rolls would normally exist in each hemisphere. When the motion is downward, the magnetic fields will be concentrated together; when the downflow zone has reached latitude 30 degrees, the fields become strong enough to float up and produce sunspots.

All this is interesting, but it is not conclusive. We cannot even be sure that the solar cycle is permanent. Between 1645 and 1715, for example, it seems to have been virtually suspended; there were few spots, and at eclipses only a very reduced corona was seen. This period is the so-called Maunder Minimum, coinciding with a cold spell when annual frost fairs were held on the frozen Thames each winter.

There is also evidence that the Sun's surface consists of many regions which are vibrating up and down with periods of around five minutes. In any one region the vertical oscillations will amount to about 15 miles; once an oscillation has started, it may persist for many cycles before dying out. These oscillations are due to seismic waves in the Sun, vaguely akin to the seismic waves produced in our world by earthquakes.

In the Sun, the surface oscillations seem to be due to the intermingling of many modes of oscillation to produce 'standing waves' which vibrate up and down in particular locations until the pattern is disturbed. The speed of the wave depends upon the temperature of the material: the higher the temperature, the greater the speed. If a wave begins by moving inward at a steep angle, it travels into hotter regions, picks up speed, and is refracted back towards the surface. When it meets the surface it bounces back, looping deep into the interior before returning to the surface and bouncing back again. If such a wave returns to its starting-point after a whole number of reflections, it will set a standing wave-pattern which may continue to vibrate for days. The fewer the number of rebounds, the deeper the wave penetrates into the Sun's interior. By analyzing the patterns of vibration, solar physicists can measure various properties of the solar globe; this is the new science of 'helioseismology'.

Many problems remain. In particular, the Sun seems to be emitting far fewer numbers of the strange elusive neutrinos than it should theoretically do, and from this it may well be concluded that the convective zone is deeper than we have thought. Moreover, it now seems that — contrary to earlier views — the rate of rotation does not increase significantly near the core. It may even be said that we know less about the Sun than we fondly imagined a few years ago!

37   Local Quasars?

Earlier in this book ('Red Shift or Red Herring?') I have said something about the doubts concerning our estimates of the distances of very remote objects: galaxies and quasars. These estimates depend upon the Doppier shifts of the lines in their spectra. A red shift indicates a speed of recession; if no other effects are involved, it seems that some of the quasars may be around 13,000 million light-years from us, racing away at over 90 per cent of the velocity of light, so that we are probing out almost to the limit of the observable universe.

For the May 1988 *Sky at Night* programme I was joined by Dr Halton C. Arp, formerly of the Mount Wilson Observatory and now of the Max Planck Institute in Munich. Arp is the champion of the theory that the red shifts in the spectra of quasars and remote galaxies are not pure Doppler effects. In his own words: 'Our estimates of the size, age and nature of the universe are all based on the key assumption that the red shifts tell us distances. If this is wrong, our whole picture falls apart.'

What Arp has done is to check on objects which are fairly obviously

DR HALTON C. ARP

associated, either by position or by visible connections such as luminous 'bridges', and yet which have totally different red shifts. There are many examples. For example, we have high-red-shift quasars close to low-red-shift disturbed galaxies, so that evidently the quasars have been ejected from the cores of the galaxies; we have quasars attached to galaxies by 'bridges', and we also find high-red-shift galaxies linked to galaxies whose red shifts are much lower.

The conventional explanation is that we are dealing with nothing more profound than line-of-sight effects, but Arp maintains that the statistical improbability of this is overwhelming — no more than one chance in a million in some cases. The 'bridges' are even more significant. They can be seen, and are also sources of radio emission. One striking example is that of the quasar known as Markarian 205, which is attached to a lower-red-shift galaxy by a bridge whose reality can hardly be questioned. There is also the famous Stephan's Quintet, made up of five galaxies, clearly associated, but of which one member has a much lower red shift than the rest. Moreover, this group is associated with a large, rather disturbed spiral some distance away, NGC 7331, which is not unlike the Andromeda Spiral. Finally, companion systems of large galaxies usually have distinctly higher red shifts than the large galaxies themselves.

According to Arp, 'There is tremendous commitment on the part of many astronomers to the assumption that red shifts always indicate distance. If this is wrong, much of their current work will be rendered obsolete.' Yet if the red shifts are not pure Doppler effects, then what are they?

Of course, Doppler shifts are present; what Arp maintains is that there is also an extra non-velocity component. He believes that new matter is born within the active nuclei of galaxies, and when created it is of intrinsically high red shift; it is expelled in the form of small, compact quasars whose red shifts decay with time as the quasars evolve into young galaxies. This means that the quasars with the highest red shifts are intrinsically the least luminous, rather than being the most powerful. It is even possible that there are quasars inside our Local Group of galaxies, and there is indeed evidence of quasar clustering associated with nearby systems such as Messier 33, the Triangulum Spiral, which is not much more than a couple of million light-years away. Quasars are also associated with the famous irregular galaxy Messier 82 in the Great Bear.

If the newly formed matter is of low mass — that is to say, its constituent protons and electrons have low mass — then its red shift will be high. Sir Fred Hoyle and Jayant Narlikar have shown that there is a rigorous physical theory (Conformal Gravity) which will permit the masses of elementary particles to grow with time. After all, P. A. M. Dirac discussed the creation of matter more than a generation ago; Hoyle, Bondi and Gold discussed it in the concept of their steady-state theory; and in 1965 Andrei Sakharov suggested spontaneous 'seeds' of galaxy creation. Allan Guth and Andrei Linde have recently declared that matter-creation is possible within the fashionable inflationary-universe theory, while Ilya Prirogine and collaborators have shown how matter can appear in the general relativity equations. According to Arp, 'It is all really like a fluctuation in the material vacuum — and the matter appears: why not?'

Arp's own 'hunch' is that the connecting bridges discussed earlier are analogous to the arms in a spiral galaxy, and represent matter which is flowing outward. One is reminded of a remark made by Sir James Jeans many years ago: 'Perhaps the spiral nebulae represent matter bounced into our universe from another universe.'

Certainly Arp is not alone; others who support his views of the unreliability of the red-shift estimates are Sir Fred Hoyle and Drs Geoffrey and Margaret Burbidge. If he is right, then we may be in for a revolution in cosmological thought greater than any since Edwin Hubble proved that the spirals are external systems rather than minor parts of our own Galaxy. But Arp's theories stand or fall upon his observational evidence, and for the moment it is very important to keep an open mind.

38 Meteor Streams

To the meteor enthusiast, August is the best month of the year. Between around 27 July and 17 August, the Earth passes through the Perseid stream, and we are always treated to a good display — particularly when, as in 1988, the Moon is out of the way during the night of August 11–12, when the shower reaches its peak.

Of course, not all meteors are members of shoals. There are also sporadic meteors, which, so far as we can tell, are lone wanderers, and may appear from any direction at any moment. Also there are many annual showers, and some have been much more spectacular than the Perseids, notably the Leonids of 1799, 1833, 1866 and 1966, and the Giacobinids of 1933. But it is the Perseids which are the most reliable.

Because the meteors in a shower are travelling through space in parallel paths, they seem to issue from one particular point in the sky, known as the radiant; that of the August meteors is in Perseus, not far from the border with Cassiopeia. The effect is much the same as that of motorway lanes to an observer standing on an overlooking bridge. Try it, and you will see that the lanes appear to 'radiate' from a point near the horizon.

Because a meteor may seem brilliant, it might be expected to be large, but this is not so; a bright meteor is probably no larger than a grain of sand. What we see is not the tiny particle itself, but the luminous effects which it produces as it plunges to its death in the upper air. It may meet the Earth at a relative speed of up to 45 miles per second, and it burns away at a height of above 40 miles, finishing its groundward journey in the form of ultra-fine 'dust'. Note, by the way, that there is no connection between a meteor and a larger, solid body which may land without being destroyed, and is then termed a meteorite. Meteorites come from the asteroid belt; meteors are cometary débris.

Comets, of course, are the most erratic members of the Solar System. Many of them have very eccentric orbits, and all are of small mass by planetary standards. The only substantial part of a comet is the nucleus, which is essentially icy. As the comet nears the Sun the ices start to boil off, and material is ejected; the comet develops a gas or ion tail, and also a dust tail which is usually curved. No two comets are exactly alike; some have no tails at all, and the short-period comets have lost so much material during their repeated returns to perihelion that most of them are faint. Some have even been known to break up and disappear. Such was the fate of Biela's Comet, which split in 1846 and has not been seen since 1852, though in 1872 and some subsequent years meteors were observed to issue from the place where the comet ought to have been.

When dust is ejected from a cometary nucleus it produces a trail, but the particles do not follow precisely the same path as the parent comet. Some move slightly closer-in to the Sun, other slightly further out, but gradually they are spread into a closed 'loop' centred on the comet's orbit. Obviously, this process takes some time. At first the particles remain close to the comet, and we see a meteor shower from them only when we happen to pass close by — that is to say, near the 'node', where the comet's orbit passes through the plane of the orbit of the Earth. There are two nodes, though one of them may be well outside the Earth's orbit and so can never produce a meteor shower. This happens, for example, with the Quadrantids, which have a short, sharp maximum in early January. And because the Quadrantid stream is regularly perturbed by the powerful pull of Jupiter, it is estimated that in no more than about 500 years the shower will have ceased

METEOR TRAIL As photographed by T. J. C. A. Moseley. The large blob is Venus, the smaller blob is Jupiter, and the star-cluster to the left of the trail is Praesepe.

altogether so far as we are concerned.

Relatively young meteor streams — that is to say, those with their particles still close to the parent comet — produce major displays only occasionally, because the Earth has to be in just the right place at just the right time. This is why the Leonids of November, associated with Comet Tempel–Tuttle, yield brilliant displays only at 33-year intervals (and even then, 1899 and 1933 were missed). We may expect another Leonid storm in 1999. The Giacobinids, associated with Comet Giacobini–Zinner, produced major displays in 1933, 1947 and 1985. Middle-aged streams have had time to spread all round the orbit, but are still fairly narrow; the Quadrantids are typical, and because the orbit of the stream 'swings around' to some extent, due to planetary perturbations, we may penetrate different parts of the stream in different years, so that the displays are by no means uniform. Older streams are more consistent, though generally less rich. The meteors associated with Halley's Comet produce showers at both nodes, the Eta Aquarids in May and the Orionids in October. At the last return of the comet, in 1986, the associated meteor streams were not noticeably richer than ususal — but Halley's Comet has been in its present orbit for a long time.

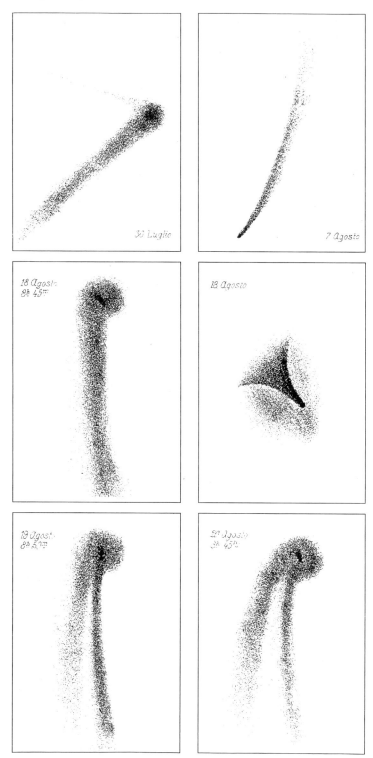

30 Luglio

7 Agosto

18 Agosto
9h 45m

18 Agosto

19 Agosto
9h 5m

20 Agosto
3h 45m

DRAWINGS OF
COMET SWIFT-
TUTTLE, MADE
IN 1862 BY G.
SCHIAPARELLI

The Perseids have their own comet: Swift-Tuttle, which was last seen in 1862, and was estimated to have a period of 120 years. It was expected to return in or about 1982, but up to now (1989) it has failed to put in an appearance. Its orbit is necessarily uncertain, and non-gravitational forces may produce extra perturbations. If, as has been suggested, it is identical with Kegler's Comet of 1737, the period could be between 129 and 130 years, and it could return around 1991–2. We must wait and see. My personal view is that it has come and gone undetected, but I may well be wrong.

Until fairly recent times, visual observers provided most of the data for meteors, including their heights; if the same meteor is recorded by two observers, separated by several miles, the apparent position in the sky will be different, and the altitude can be worked out by relatively simple trigonometry — a method pioneered in the 1790s by two German students named Brandes and Benzenberg. Today radar is widely used, since a meteor trail will reflect a radar pulse in much the same way that a solid body will do. Amateurs can make useful contributions. In Sussex Dr John Mason has built up a receiver which enables him to pick up brief snatches of conversation from a broadcasting station in Poland. Each 'transmission' lasts for only a few seconds at most, but some words are clearly audible, and I am sure that if I could speak Polish I would be able to recognize them.

Meanwhile, the Perseids are still with us to delight us each August. They will be seen for many centuries to come, and it would indeed be sad to lose our main annual display of cosmic pyrotechnics.

---

39 Watch this Space!

Look up into a clear night sky, and you will see vast numbers of stars. If you use optical equipment you can also see many external star-systems or galaxies. But what about the space between the stars? Is it empty, as we used to think? The answer is a resounding 'no'.

The first direct proof of the existence of interstellar material was obtained in 1904 by a German astronomer named Johannes Franz Hartmann. He was looking at the spectrum of a certain star, Mintaka or Delta Orionis, the northernmost star of Orion's belt. It is a binary, made up of two components so close together that telescopically they cannot be separated, but both yield visible spectra. They are in orbit round their common centre of gravity, and this causes shifts in the

FRAUNHOFER LINES IN SPECTRUM

positions of the lines of the two stars. According to the Doppler effect, the lines are blue-shifted when a star is approaching, red-shifted when it is receding, so that with Mintaka the lines appear to oscillate round a mean position. (Of course, these effects are superimposed on the overall radial velocity of the system, but the principle is clear enough.)

However, Hartmann found that there were some dark lines in the spectrum which did not take part in the oscillation. Obviously, then, these lines were due to material lying between the star and ourselves — in fact, to interstellar material (in this case, calcium). Space contains thinly-spread gas, and also what we may term 'dust'.

Quite apart from its imprints on the spectrum, the dust has another effect also: it causes reddening — which will be appreciated by anyone who has watched a blood-red setting Sun shining through the thick, dusty layers of the Earth's atmosphere. The true colour of a star can be determined from its spectrum. White stars are hotter than yellow, yellow hotter than red. If a star which we know to be hot and white looks reddish, than we can be sure that its light is coming to us via interstellar dust.

Actually, dust is only a minor constituent in the interstellar medium, but we know of regions which are well above average density in both dust and gas. There are small, well-defined 'clouds' known as Bok Globules (in honour of the Dutch astronomer Bart Bok, who first drew attention to them) which may be embryo stars. We can also see reflection nebulae, which shine by the light of stars in and near them; the lovely nebula near Rho Ophiuchi is a classic example.

Probably the most famous of all nebulae is Messier 42, in the Sword of Orion. This is what is termed an H.II region; the interstellar matter is lit up in the manner of a fluorescent lamp by new, hot stars which have formed out of the condensing gas-clouds, so that the nebular material is emitting a certain amount of light on its own account. However, nebulae such as Messier 42 are only parts of much larger clouds which are harder to detect, but which betray their presence by molecular lines in their spectra — mainly in the radio range. There are also clouds of cold hydrogen which emit at a wavelength of 21 centimetres, and enable us to map the shape of our Galaxy. The distribution of these hydrogen clouds shows up the forms of the spiral arms.

What do we know about the origin of the interstellar material? It seems that the gas out of which the Galaxy was formed, perhaps 15,000 million years ago, consisted only of the light elements hydrogen, helium, some deuterium ('heavy hydrogen') and so on, which existed soon after the Big Bang. It was only after stars were formed out of this condensing material that heavy elements were produced, through the process of nucleosynthesis in the stars' cores. Once a star has been formed there is a lengthy static period, but eventually the star will start to shed its material. It may do so gently, producing a planetary nebula (as our modest Sun will do) or, if more initially massive, it may explode as a supernova.

Supernovae are of supreme importance, because they return material to the interstellar medium. The original stars processed it, and created heavy elements out of the primordial hydrogen and helium, so that when a supernova explodes it enriches the interstellar medium with these heavy elements.

There is a range of densities in the interstellar medium, but most of the material is very tenuous and at a very high temperature. (Bear in mind that the scientific definition of 'temperature' is not the same as what we ordinarily mean by 'heat'. Consider a firework sparkler together with a red-hot poker. Every spark of the firework is 'white-hot', but contains so little mass that there is no danger in holding the sparkler in one's hand, whereas I for one would be most reluctant to grasp the lower-temperature end of the glowing poker.) Supernovae eject their high-temperature material violently, producing what may be called 'bubbles' of gas throughout the Galaxy; in fact, only about two million years is taken for most of the volume of the Galaxy to be filled with coalescing supernova shells. We may describe it as a vast high-temperature 'soapy foam', with interspersed cool clouds. The gas in the Galaxy is contained mainly in the disk, but there is also a tenous halo which has been enriched by the supernova explosions.

Everything is in a state of flux. The cooler, denser clouds of interstellar material tend to be swept into pancakes by the expanding shells, and it is out of this material that stars form. Our Galaxy is far from static. It is continually in turmoil; wherever you are, sooner or later the searing blast of a supernova shock-wave will come past you.

We can also see interstellar material in other galaxies; a striking example is the so-called Sombrero Hat Galaxy with its dark band, and the masses of dark material in the comparatively nearby radio Galaxy Centaurus A. Moreover, we can detect the presence of intergalactic material because of its imprint on the spectra of quasars, which,

RHO OPHIUCHI AND THE REGION OF ANTARES Gas and dust clouds reflecting light; the red supergiant Antares; the yellowish globular star cluster M4, at 5700 light-years from the Sun. A colour photograph was produced by combining three separate black-and-white pictures taken through different colour filters with the UK Schmidt Telescope.